NOV -- 2013

Scenic Maine
ROAD TRIPS

Scenic Maine
ROAD TRIPS

Dan Tobyne

MAINE

ISBN 978-1-60893-220-7

Cover design by Lynda Chilton
Interior design by Jennifer Anderson

Printed in China
5   4   3   2   1

Down East

Distributed to the trade by National Book Network

Library of Congress Cataloging-in-Publication Data available

This book is dedicated to my parents,
Roy and Louise, and the special person
who once told me, "You need to begin
by knowing you've already arrived."

Contents

Introduction   8

1. Pequawket Trail   11
2. Grafton Notch   21
3. Rangeley Lakes   33
4. Old Canada Road   43
5. Route 27   55
6. Moosehead Lake   65
7. Katahdin Woods & Waters   77
8. Acadia All-American Road   87
9. Schoodic National Scenic Byway   101
10. Black Woods Scenic Byway   111
11. The Million-Dollar View   121
12. The Bold Coast   127
13. Fish River   141
14. Saint John Valley   151

Acknowledgments   160

Introduction

MAINE, THE LARGEST OF THE NEW ENGLAND STATES, IS TUCKED
away in the extreme northeast corner of the country and
is the only state in the nation bordered by only one other
state. Some people believe this distance from everyone
else is one of the reasons Maine has such a strong cultural
heritage, not to mention a reputation for being somewhat
curmudgeonly. It is, however, definitely one of the things
that make it a unique and interesting place to visit.

Henry David Thoreau often traveled to Maine. He understood that Maine was special and although he may not
have coined the phrase "The way life should be," he did
say, "What a place to live, what a place to die and be bur-
ied in! There, certainly, men would live forever . . ." Today
anyone can follow Thoreau's footsteps and visit many of
the places he wrote about by exploring the Moosehead
Lake Scenic Byway, one of fourteen designated scenic
roadways in the state.

Maine's byway program is one of the oldest in the
country, dating back to the early 1970s. It connects travelers to the culture and legacies of the past, as well as to
the customs and defining character that is Maine today.
These views and landscapes offer insights into the intrinsic
values of the people of Maine. There, we can walk in the
footsteps of many others: those who came to look for
adventure, build fortunes, invent things, fight wars, or sim-
ply call this place their home. From the Down East coast to
the western mountains and the northern plain, Maine offers
a diverse collection of educational and recreational experiences. The state is home to the highest mountain on the east
coast of North America, and it's the first place to feel the
rays of the rising sun at the beginning of each day. It's also
home to the last great wilderness area in the eastern United
States, the seventeen-million-acre Great North Woods. The

first real gold rush in America occurred in Maine and to-day you can still successfully pan for gold. Benedict Arnold marched through the state on his ill-fated attempt to capture Quebec during the Revolutionary War. With its fertile soil, the northern plain is famous for Maine potatoes and Acadian culture, and the state is known all over the world for its famous lobster. All this and much more can be found on the roads and byways of Maine.

It wasn't until 1991 that the National Scenic Byways program was created to recognize roads that have contributed significantly to the American experience. Five of Maine's byways have been recognized at the national level, with one designated an All-American Road. These are scenic on a countrywide scale, and each offers a unique perspective. Visitors come from all over the world to enjoy these vistas.

To be considered a National Scenic Byway a roadway must first be designated at the state level, be sponsored by an organization, and adequately demonstrate at least one of the following attributes beyond scenic beauty: archaeological, cultural, historic, natural, or recreational significance. An All-American Road must excel in at least two of these areas. These fourteen byways offer a unique traveling experience, and they are truly roads into the heart and soul of Maine.

— *Dan Tobyne*

Pequawket Trail

1

PEQUAWKET TRAIL

REGION: Lakes & Mountains
LENGTH: 60 miles
ROADS: Routes 5 & 13

TRAVEL TIME: 2 hours
START: Standish
END: Gilead

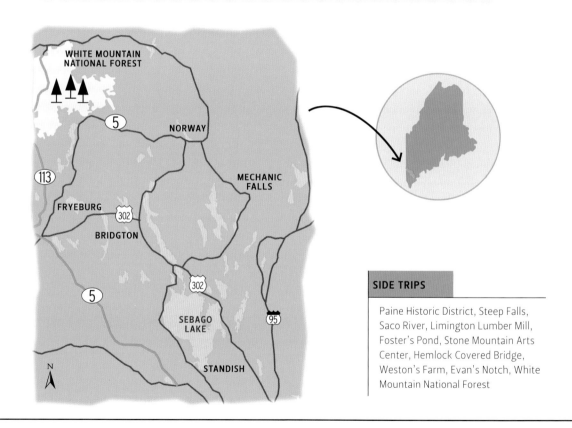

SIDE TRIPS

Paine Historic District, Steep Falls, Saco River, Limington Lumber Mill, Foster's Pond, Stone Mountain Arts Center, Hemlock Covered Bridge, Weston's Farm, Evan's Notch, White Mountain National Forest

Previous page: The Androscoggin River near Gilead. Right: North Fryeburg Farm. Below: Dandelions in the Saco River Valley.

THE PEQUAWKET TRAIL IS LOCATED IN CUMBERLAND AND OXFORD counties and runs north through the Saco River Valley from Standish to Gilead. The sixty-mile-long route incorporates numerous small towns and villages along the way and even crosses briefly into New Hampshire. The lower portion of the byway, from Standish to Fryeburg, follows the historic route of the Sokokis Indians (a branch of the Eastern Abenaki), who used the trail to move between their summer encampments on the coast and their traditional homes located in the area they called Pequawket, present day Fryeburg.

Formerly known as Pearsontown Plantation, the town of Standish was incorporated in 1785. Its first European settlers were Massachusetts veterans of the French and Indian War. Land was often given as payment to soldiers for their service and at the time Maine was still part of the Commonwealth of Massachusetts. Baldwin (Flintstown Plantation) and Brownfield (Brownfield Plantation) also have their roots in land grants to former soldiers. Some veterans sold their land rights, but many brought their families and began to work the land. These early settlers usually built along the trail for

Corn harvesting along the West Fryeburg Road.

convenience and also for protection. When Europeans began colonizing this particular part of Maine it was not wilderness. Maine's indigenous population of Sokokis had lived in the area for thousands of years and had developed the region as a trade route with the French in Canada, so hostilities often broke out between the Sokokis and the English settlers. Smallpox, land grants to former soldiers, and the township grant for the land around what is today Fryeburg combined to slowly minimize the Sokokis as a factor in the development of the region. The last Indian raid in New England occurred in the upper Saco River Valley in 1781. A small band of Indians attacked homes in Newry, Bethel, and Gilead. Some residents were killed and some captured before the Indians fled to Canada, ending the existence of the Sokokis as a tribe in the region.

Beginning at the junction of the "Two Trails" in Standish, this trip quickly exposes travelers to a slower and less hurried pace, a "turning back the clock" pace, an experience virtually forgotten in today's busy world. Along the way you'll begin to notice small groups of homes and buildings interspersed between rolling hills of open farmland and woodlands.

A good example of this type of settlement is the Paine Historic District, located just north of Standish Center. This small cluster of homes on the eastern side of Watchic Pond includes a cemetery enclosed with a white picket fence and shaded by large maple trees. The cemetery is filled with the names of the Paine, Mayo, and Higgins families. These multi-generational cemeteries often have a story to tell and can be interesting to investigate. In the eighteenth and nineteenth centuries, water power was the main form of energy for industry and because the Saco River watershed had an abundance of this natural resource, the villages along the trail were soon home to numerous lumber and woolen mills. Steep Falls, another village in Standish, was a mill town with various pulp and lumber mills located along its riverbank. Visitors who stop here today can see the remains

of the last pulp mill located at the falls for which the town was named.

As industry made inroads into the community, so did the railroad. The railroad lines that cross back and forth over the byway belong to the Mountain Division Railroad. The Portland & Ogdensburg Railroad was supposed to connect Portland, Maine, to Ogdensburg, New York. Sections of that railroad were never built, but a line did run from Portland to Sebago Lake and between many of the towns in the region. Over time, the Maine portion of the P&O became the Mountain Division Railroad. Currently not in use as an active railroad, some sections have been converted to Rails with Trails. This type of trail system runs next to the tracks but not on the track bed itself. The plan is to create fifty-two miles of trails connecting Portland to Fryeburg.

Baldwin and Hiram are the next towns on this trip. Route 113 and the railroad both follow the Saco River in this section and you'll be rewarded with intermittent views of the river and the White Mountains. Two points of interest are the historic Brown Memorial Library, home to the Baldwin

Historical Society, and the Limington Lumber Company Mill. With its proximity to the road, vast yards of white pine logs neatly piled and waiting to be milled, and thousands of board feet of cut and stacked lumber, the Limington mill offers a great view of an industry that once dominated the woods of Maine.

In Hiram the river begins to twist and turn, creating numerous small ponds and bogs. Ingalls Pond is one example. Also known as Foster's Pond, it's considered one of the better smallmouth bass ponds in the region.

Continuing on, you enter the town of Brownfield. At this point, you'll notice a change in the nature of the surrounding area — due to the 1947 fire that consumed most of Brownfield, including more than 85 percent of its historic buildings. However, and maybe as a result of the Great Fire, this stretch provides wonderful panoramic views of the White Mountains, including Frost, Stone, and Burnt Meadow mountains. Brownfield is also home to the Stone Mountain Arts Center, a unique concert venue that hosts nationally recognized artists in this beautiful part of Maine.

Leaving Brownfield, the byway enters Fryeburg, where the Saco River splits. In the 1800s a six-mile-long canal was built that effectively cut off fifteen miles of the "Old Course" of the river. The entrance to this part of the river has slowly filled with silt and today the canal is considered the official river. Fryeburg is a historic village with many registered buildings and it is home to Fryeburg Academy. The Fryeburg Fair is Maine's biggest fair and hosts more than 300,000 visitors each September. Fryeburg is also the second busiest entrance to Maine and has become a favorite destination for visitors on day and weekend trips. The area is home to antiques shops, historic inns, and outdoor sports shops

Left: East Fryeburg farm. Right: Antiques shop near Gilead. Below: Hemlock Covered Bridge.

specializing in canoe and kayak rentals. You can also see the Hemlock Covered Bridge, a 109-foot truss bridge spanning the "Old Course." To see the bridge other than from the river, you need to find Frog Alley off Route 5 and continue to the end, where you'll be rewarded with a classic example of a truss-constructed covered bridge.

After leaving downtown Fryeburg, your route follows River Road and passes historic Weston's Farm, a working farm that includes an interesting farm stand selling fresh produce and a variety of local products. Just past the farm you cross the Saco River and enter the flat valley floor opening up to panoramic views of farmland — mostly potatoes and corn, back-dropped by the majesty of the White Mountains.

The northern portion of the byway begins in this flat and fertile bottomland, but beyond Stow, a little village located at a curving slope in the road, there's a visible shift in

topography. Farmland begins to fall away and arable land is replaced by hilly wooded terrain. The byway leaves Maine and enters New Hampshire, but after only a short distance a large sign welcomes you back to Maine and you enter Maine's portion of the White Mountain National Forest.

This section of the national forest is comprised of approximately 47,000 acres of remote mountainous forest and includes the Caribou-Speckled Mountain Wilderness. There are primitive campgrounds, hiking trails, and picnic areas, but very little additional development, leaving the traveler with the correct perception that they're traveling through a truly wild place. The byway weaves its way toward Evans Notch, following the Cold River. The notch's majestic views of the glacially formed terrain are inspiring, especially in the fall, but if you have a real desire to see the view in winter, be prepared for some snowshoeing, as this section of the road is not plowed in winter. From this point the byway follows Evans Brook as it flows down the mountain, meeting up with the Wild River and eventually the Androscoggin at Gilead.

2

Grafton Notch

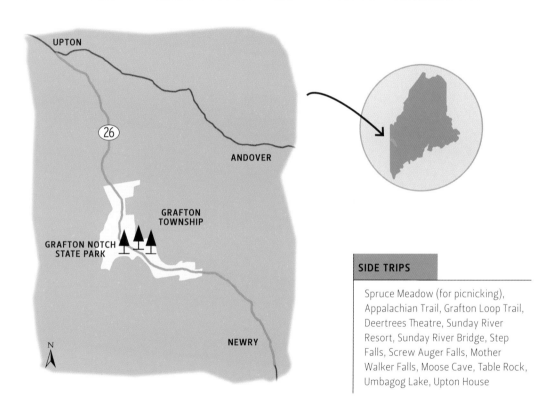

2

GRAFTON NOTCH

REGION: Lakes & Mountains
LENGTH: 21 miles
ROADS: Route 26

TRAVEL TIME: 2 hours
START: Newry
END: Upton

UPTON

26

ANDOVER

GRAFTON
TOWNSHIP

GRAFTON NOTCH
STATE PARK

NEWRY

N

SIDE TRIPS

Spruce Meadow (for picnicking), Appalachian Trail, Grafton Loop Trail, Deertrees Theatre, Sunday River Resort, Sunday River Bridge, Step Falls, Screw Auger Falls, Mother Walker Falls, Moose Cave, Table Rock, Umbagog Lake, Upton House

THE GRAFTON NOTCH SCENIC BYWAY BEGINS IN NEWRY TO THE south and runs twenty-one miles to its terminus in the small northern town of Upton. Along the way, you'll discover amazing geological features like Screw Auger Falls, the Appalachian Trail, and the lost town of Grafton. It offers a wealth of outdoor activities, including rock and mountain climbing, snowshoeing, cross-country skiing, hiking, camping, and bird watching. It also offers less traditional activities such as prospecting and geocaching.

Before starting up the notch, it's worth noting the town of Bethel, which borders Newry. Once known as Sudbury-Canada Plantation, Bethel, was incorporated in 1796. The railroad came soon after, bringing tourists and

Previous page: Bear River, Grafton Notch. Above: Sunday River Bridge, Newry.

commerce; and with the completion of the commerce road from Portland to Errol, New Hampshire, the town began to grow. Today, Bethel is considered one of western Maine's quintessential village towns, and along with Newry, it's home to the Sunday River Ski Resort, a four-season recreation area and one of the premier ski resorts in the east. It's also home to the Bethel Inn, a truly grand resort hotel with a colorful past.

Consider taking a short detour to visit one of Maine's nine covered bridges. The Sunday River Bridge is an 87-foot-long Paddleford truss bridge, and is the most painted and photographed covered bridge in Maine, earning it the nickname, the Artist's Covered Bridge. To get to the bridge, turn left off Route 26 onto Sunday River Road. It's located about a mile and a half down on the left, at a quiet bend in the Sunday River. The bridge is accessible to foot traffic but has been closed to vehicles since 1958.

The official starting point of the Grafton Notch Scenic Byway is in the town of Newry. Once known as Bostwick Plantation, it was renamed by immigrants from Newry, Ireland. Taking advantage of the fertile soil of the Bear River Valley and well-traveled road, the inhabitants established themselves as a farming community. Hay was the major crop and for many years it supplied the lumbering enterprises that were developing in the woods surrounding Grafton Notch. Traveling this route today you can still see active farms as well as old barns and farm buildings no longer in use, vestiges of an earlier time.

Newry's bottomland follows the Bear River's winding path along the valley floor. The Bear River is one of twenty-two major tributaries of the Androscoggin River, fourteen of which are in Maine. The Androscoggin, along with its tributaries, were part of the timber transport system used in the nineteenth and twentieth centuries that brought logs downriver to numerous pulp, paper, and lumber mills throughout Maine.

Below: Route 26 near the top of Grafton Notch. Right: Old hay barn in Newry.

The old Grange Hall, roadside stands, and wild apple trees fall away in the rearview mirror as the road begins to climb up to Grafton Notch State Park. The park is named after one of the lost towns of Maine. Born and built around timbering, Grafton was incorporated in 1852, but once the trees were gone, the town followed. Like many places in Maine, much of the land was owned by speculators hoping to get rich by harvesting trees. The town was located above and below the notch and lumbering allowed its inhabitants to work year-round, logging in the winter and farming the rest of the year. When the trees were gone the citizens eventually moved on and what remained was bought by the Brown Paper Company and razed. Brown reforested the land and today there is little evidence a town ever existed. Once through the valley you begin to see patches of birch and fir, as well as other hardwoods transitioning back to primeval forest.

Geologically, the notch is part of Maine's Mahoosuc Range and is an example of intense glacial activity as well as more recent erosion. You don't need to go very far to see a wonderful example of this kind of activity. About seven miles into the journey you will come to a parking lot on the right side of the road that is part of the Nature Conservancy land at Step Falls. This set of stepped falls includes numerous pools that offer swimming and wading opportunities during lower water levels. A short walk into the woods will bring you to the falls. Farther down the road and not long after entering the state park you will come to a parking lot on the left for visitors to Screw Auger Falls. The first thing

Left: Screw Auger Falls, Grafton Notch State Park. Right: Trail near Mother Walker Falls.

you will notice is a small waterfall that spills onto a large granite slab. You will also see a few glacial potholes worn out of the granite stone. The water traverses the slab and enters a screw-shaped set of cutouts before cascading over a ledge and dropping twenty feet into a small pool. The stone sides and bottom of this section are smooth and rounded, with lots of curves and holes that in one case has created a natural bridge, something rarely seen in New England. There are picnic tables and Screw Auger Falls is a pleasant place for a break — as long as it's not blackfly season.

Another mile farther up the road you will come to a small parking area for Mother Walker Falls and Moose Cave. The falls were named after a local resident who lived in Grafton, and the cave got its name after a man discovered a moose that had fallen and was stranded in the cave. The moose ended up on the man's dinner table.

Mother Walker Falls is a 980-foot-long flume that drops about one hundred feet as it steps down through a narrow gorge. It's very impressive when the river is running high, but can be dangerous, so watch your footing. There's a loop trail that will take you to Moose Cave — you should pick up a map at the entrance to the park. Another loop trail that's popular is the Table Rock Trail. This loop is only 2.5 miles long from the parking area, but the terrain is difficult in places and requires a certain level of fitness. The destination is Table Rock, part of a large cave system created by large slabs that have combined to create an intriguing array of holes and caverns. Table Rock affords great views of Old Speck Mountain and the notch for those willing to make the climb.

There are possibilities along this section for all levels of hikers, and some will test the best adventure-bound trekkers. The Appalachian Trail (A.T.) crosses at the Table Rock parking area, so anyone who hikes the loop trail can say they have hiked part of the A.T., but if you're looking for a challenge you should try the Mahoosuc Notch Loop.

View from the Appalachian Trail in the Mahoosuc region. (Jonathan Hines)

Ninety-five percent of this trail is located in Maine, but the parking lot is actually in New Hampshire. The loop is 6.5 miles, round-trip, and starts from a parking lot located on Success Pond Road, just north of Berlin, New Hampshire. Don't let the relatively short length of the hike fool you. Many believe the Mahoosuc Notch is the most difficult section of the entire Appalachian Trail. The halfway point of the hike is at Speck Pond, and if you decide to hike the loop in a counterclockwise rotation, you might want to rest at the pond's campsite because this means you've hiked — or climbed, as some people would say — "the toughest mile of the Appalachian Trail." If you really want a challenge, hike back again the way you came. At this point, however, most hikers will find the Speck Pond Trail to be a better route back.

The newest trail, completed in 2003, is known as the Grafton Loop Trail. This 39-mile loop can be accessed from two locations: the Loop Trail parking lot at Newry in the south and the Table Rock Appalachian Trail parking lot to the north. The trail has both western and eastern loops and covers some of the most spectacular landscape Maine has to offer.

Once over the notch, you'll pass through the upper section of old Grafton. If you look closely you can still see a few old stone walls and parts of the former roadway, but little else remains. You are now traveling within the Cambridge River Valley and entering the town of Upton. With a population of less than one hundred, this sleepy little town that once had its roots in the logging industry is today known more for outdoor sports opportunities. Upton is bordered in the north by Umbagog Lake, one of the premier fishing destinations in New England. The town is also a hotspot for snowmobiling in winter and hiking and mountain biking in the summer. It's also home to the Upton House, a renovated 1886 farmhouse that offers the perfect location for a country inn getaway in any season.

Old farm and wild apple orchard in the Bear River Valley.

3

Rangeley Lakes

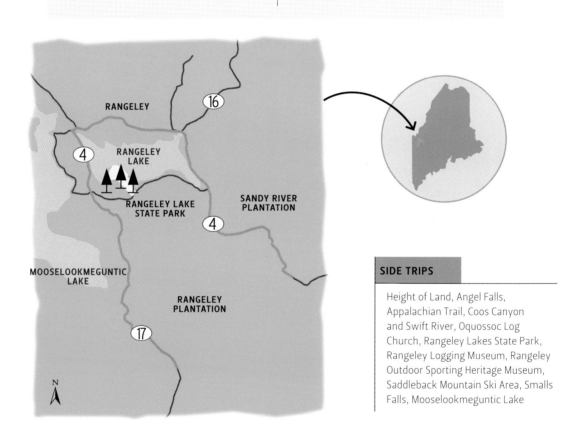

3

RANGELEY LAKES

REGION: Lakes & Mountains
LENGTH: 52 miles
ROADS: Routes 17 & 4

TRAVEL TIME: 2.5 hours
START: Byron
END: Madrid Township

RANGELEY

RANGELEY
LAKE

RANGELEY LAKE
STATE PARK

SANDY RIVER
PLANTATION

MOOSELOOKMEGUNTIC
LAKE

RANGELEY
PLANTATION

N

SIDE TRIPS

Height of Land, Angel Falls,
Appalachian Trail, Coos Canyon
and Swift River, Oquossoc Log
Church, Rangeley Lakes State Park,
Rangeley Logging Museum, Rangeley
Outdoor Sporting Heritage Museum,
Saddleback Mountain Ski Area, Smalls
Falls, Mooselookmeguntic Lake

Right: View toward
Rangeley Lake from
scenic overlook.

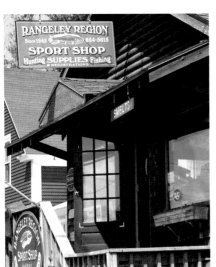

THE RANGELEY LAKES REGION OFFERS ONE OF FOUR NATIONALLY recognized byways in Maine and it features a unique set of natural resources. Lakes, mountains, rivers and streams, working forestland, and an abundance of wildlife combine to make this special place an outdoor sporting paradise. Today, the Rangeley area is a four-season destination and has something for everyone who loves being outside. The fishing in this region is legendary and, along with hunting and camping, is one of the first activities that attracted visitors to the area.

The region is made up of three distinct watersheds: the Kennebago, Mooselookmeguntic, and Magalloway, and much of it today remains as it was when first explored by Europeans more than 250 years ago. Europeans, however, were not the first to visit this heavily forested land. There's evidence that native inhabitants as far back as 11,000 BC traveled to this part of Maine hunting caribou and other game. A number of archeological digs in the area have turned up a treasure trove of artifacts.

This trip follows Routes 17 and 4, passing through the village of Oquossoc, the town of Rangeley, and Madrid Township as it traces a horseshoe shape on the map. Beginning just beyond the town of Byron, home of famous Coos Canyon and the Swift River, where gold was discov-

Lower Richardson Lake at sunrise.

ered in the 1840s, the byway begins to rise in elevation as it follows the Swift River along the western ridgeline. You're heading toward the summit of Spruce Mountain and the spectacular overlook known as Height of Land. The view from here is spectacular and looking west you can see Mooselookmeguntic Lake, as well as part of Upper Richardson Lake (Molechunkemunk). You also cross the Appalachian Trail at this point. Maine is home to 281 miles of the A.T. and considered by most to be the toughest stretch among the fourteen states the trail passes through. Height of Land is located just off the summit of Spruce Mountain and just above the A.T. crossing.

You should be on the lookout for one or more moose standing by the side of the road in search of salt or just deciding it's time to cross the road. In the spring it's not uncommon to see a mother with one, sometimes two, calves in tow.

After Height of Land the road continues to follow the ridgeline and moves to the eastern slope, rewarding you with a grand view of Rangeley Lake and the hills and mountains to the north and east before starting its downward slope to Oquossoc, a quaint village within the town of Rangeley located on the shores of Mooselookmeguntic and Rangeley lakes. Oquossoc has a variety of amenities — restaurants, a grocery store, a marina, antiques shops, museum, and sporting camps are all clustered in this tiny enclave that seems to have something for everyone. There's also a service station, sporting goods store, and fish hatchery, as well as a few historic landmarks such as the famous Oquossoc Log Church, built in 1916 and added to the National Register of Historic Places in 1984.

The abundance of trophy fish swimming in the local lakes and rivers was one of the best-kept secrets of the mid-1800s, known only to a few adventurous types who made the trek to this out of the way part of Maine. These individuals were known locally as "sports." Travel to the

region usually included a train ride, a wagon ride, and then a good amount of walking. Some sports walked as many as ten miles to reach their final destination. At some point in the mid-1800s steamboats began to run on the lakes, making the journey more practical, but easy access didn't come to the region until 1890, when the 28-mile Phillips and Rangeley Railroad began operations and travelers could ride undisturbed to their destination. Once others discovered this fishing mecca, the region was flooded with sportsmen and women of all ages and a new commercial enterprise was born.

Sporting camps, large hotels, and casinos began to spring up everywhere, especially at the edge of the larger lakes. Today, one of the last remaining historic inns in the western lakes region is the Rangeley Inn. Standing on Main Street in downtown Rangeley, the inn was built in 1897 and accommodated wealthy guests from all over the East Coast. Sporting camps were a different story. They were built for the adventurer who came to the region in search of trophy fish and game. These were rustic enterprises and often employed guides to assist the anglers and hunters in pursuit of their goals.

Left: Lakewood Camps on Lower Richardson Lake. Below: Flyfishing the Rapid River. Right: View from the shore, Lakewood Camps.

Fishing was so good in the early years that brook trout and salmon weighing more than eight pounds, with some as large as twelve pounds, were the norm and not the exception. On a good day anglers could catch as many as twenty fish apiece; so many that they didn't know what to do with them all, and often just discarded them after the customary photograph was taken. This gave rise to the first conservation groups in New England, formed to foster sustainability of the regional fishing stock.

With the development of alpine skiing and the increasing popularity of snowmobiles, the area has become a destination for winter activities as well. Saddleback Mountain,

located just outside town, was not the first location for skiing in the area. One of the first locations was at Rangeley Manor in the 1930s and later at Hunter's Cove, but today Saddleback Mountain is the ski resort of choice in this part of New England. It is the third largest ski area in Maine and can accommodate all levels of expertise.

Rangeley is also home to one of the largest snowmobile clubs in Maine as well as hundreds of miles of groomed trails. And it hosts one of the biggest winter events in Maine, the Rangeley Lakes Region Snowmobile Snodeo, a four-day festival with a large roster of on- and off-trail activities that attracts people from all over the world. Another growing winter activity is the New England Pond Hockey Festival, which is held in early February on Haley Pond, located in downtown Rangeley.

Leaving Rangeley you'll soon pass Long Pond, also called Beaver Mountain Lake, on the right-hand side of the road. The road meanders along the elevated side of the pond and begins moving downward in elevation. As Long Pond folds into your rear view you'll discover another body of water on your left. This is Sandy River Ponds, part of the Sandy River, which you'll soon see cascading beside the road. Then you'll cross the Appalachian Trail again on its way to the summit of Saddleback Mountain and onward to the Bigelow Range. A final treat as you approach Madrid and the end of the Rangeley National Scenic Byway, is 54-foot Smalls Falls, a scenic waterfall with a colorful gorge, excellent swimming holes, and, if you thought to pack a lunch, a picnic area.

Left: Saddleback Mountain. Right: Coos Canyon and Swift River, Byron.

4

Old Canada Road

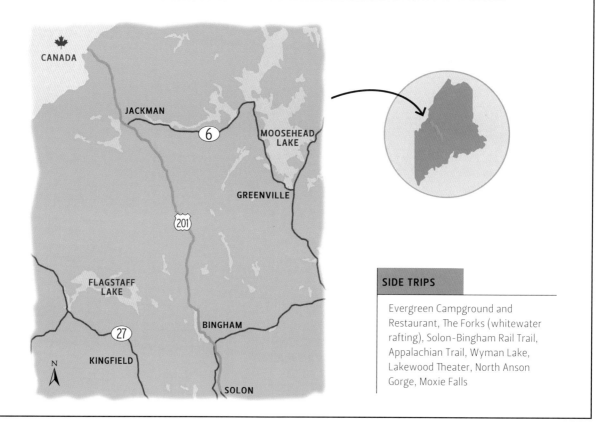

4

OLD CANADA ROAD

REGION: Kennebec & Moose River Valley
LENGTH: 78 miles
ROADS: Route 201

TRAVEL TIME: 3 hours
START: Solon
END: Sandy Bay Township

SIDE TRIPS

Evergreen Campground and Restaurant, The Forks (whitewater rafting), Solon-Bingham Rail Trail, Appalachian Trail, Wyman Lake, Lakewood Theater, North Anson Gorge, Moxie Falls

ONE OF FOUR NATIONALLY RECOGNIZED BYWAYS IN MAINE, THE
78-mile-long Old Canada Road begins in Solon and passes
through the towns of Bingham, Moscow, Caratunk, and
Jackman, as well as the plantation known as The Forks.
Strikingly beautiful, this byway has something to offer
in all seasons. Spring and summer offer a wide range
of activities, including whitewater rafting, antiquing,
canoeing, camping, hiking, and fishing. The fall brings
all the colors of foliage alive and is a favorite for leaf peep-
ers as well as a prime destination for hunters. Winter
brings skiing, snowmobiling, dogsled racing, and ice
fishing.

The Old Canada Road is part of the larger Kennebec-
Chaudiere Heritage Corridor, which began as a primitive
path through the wilderness and developed into one of the
major trade routes between southern Canada and post-
colonial America. Native Americans first used the trail to
travel through the region as they moved up and down the

Moose River Valley, and early French traders moved goods along the northern portion, creating a lucrative fur trade. Both the French (from the north) and the English (from the south) explored and settled in the region, creating conflict during colonial times, which wasn't officially settled until the end of the French and Indian War.

The extreme northern and southern sections of what is known today as the Kennebec-Chaudiere International Corridor were fairly well developed during colonial times, but the midsection that encompasses the Old Canada Road was still an unmapped and dangerous wilderness, as Benedict Arnold found out during the American Revolution, when he tried using this pathway to march to Canada to capture Quebec. He began with one thousand men and reached Quebec with fewer than seven hundred, many of whom were not fit for duty.

The byway begins at the Robbins Hill Scenic Area in Solon, a rest stop with picnic tables and an information center. The view here is inspirational, with a field of wildflowers highlighting the expanse of western mountains. A rustic farm across the road that still bales hay the old fashion way only adds to the charm.

The region has a rich history and before leaving the town of Solon a side trip is highly recommended. At Solon's village center turn left onto 201A, locally known as Ferry Street. The road takes you a short distance to the Kennebec River. Located on the left-hand side of the road, just before the Solon-Embden Bridge, is the Evergreen Campground and Restaurant. The Third Division of Benedict Arnold's Expedition camped at this site during its march to Quebec and it is the former location of one of the oldest Indian villages in Maine. Archeological explorations have found thousands of Native American artifacts and many are on display at the campground. Directly across from the campsite on the western bank of the Kennebec River is a "picture rock," known as the Embden Petroglyphs. A thirty-foot section of bedrock

Left: Wyman Lake rest stop and overlook. Below: Rolled hay bales along the Old Canada Road.

contains more than one hundred pictographs. Most likely created by the Eastern Abenaki, the petroglyphs are considered by some to be the best example of native pictographs in the eastern United States.

Once you've explored this amazing site, return to Route 201A and cross the Solon-Embden Bridge. Bear left and continue on 201A until you reach the village of North Anson. Here the Carrabassett and Kennebec rivers converge at a place called Savage Island. This section of the Kennebec River must have been a very interesting place during the days of the log drives. It is made up of numerous channels and dead ends knotted together to create a twisting field of turns and weird currents that must have made it a dangerous place for river drivers. There is also a fantastic view of North Anson Gorge on the western side of the North Anson bridge that spans the Carrabassett River. Very wide but shallow, the gorge is made up of slabs of sandstone and slate, discernible by their varying shades of gray.

Head north out of Solon on 201 to the towns of Bingham and Moscow. Bingham is located on the 45th parallel — halfway between the North Pole and the equator — and was a stop on the Somerset Railroad, which serviced the lumber industry and the towns of the Kennebec River Valley, and carried rusticators from Boston and New York to the resort at Kineo in the Moosehead Lake region. The railroad's massive Gulf Stream Trestle was located in Bingham — a marvelous piece of engineering that at 115 feet in height and 600 feet in length was one of the largest trestles in New England. The trestle is gone now but the granite abutments remain.

Just north of Bingham is the town of Moscow, home of the Wyman Dam. Constructed between 1928 and 1931, the dam generates nearly eighty megawatts of electricity. The dam created one of the gems of the Old Canada Road Byway, the beautiful twelve-mile-long Wyman Lake. The formation of the lake also flooded part of the origi-

North Anson Gorge seen from the bridge.

nal section of the roadway from Moscow to Caratunk, requiring a new section of road to be constructed. The Central Maine Power Company built the new road that now runs alongside the lake and features a number of spectacular views and public access points, including a public boat launch in Moscow and a rest area farther along with picnic tables, informational panels, and a great view of the lake.

Farther north in Caratunk, the 2,160-mile-long Appalachian Trail crosses the road and offers hikers a chance to test their skill and endurance any time of the year. Only 280 miles of the AT is located in Maine, but the adage goes: "When you cross the Maine border 90 percent of the trail is behind you and 90 percent of the difficult climbs are ahead of you."

After leaving Caratunk, the route begins to take on a very different feel. Until the mid-1970s, most of the area north of Caratunk seemed stuck in a different time. Traffic passing through was always headed someplace else and life revolved around the slowing lumber industry. All that changed in 1976 with the introduction of whitewater rafting. Small at first, the rafting business took off and today numerous rafting companies accommodate more than 80,000 adventure seekers each season.

Although whitewater rafting is now a big business, The Forks still maintains the look and feel of a quiet outpost alongside a wilderness roadway. The Forks is also home to Moxie Falls and Lake Moxie. Moxie Falls is one of the tallest waterfalls in Maine, dropping in one spot nearly one hundred unobstructed feet. To find the falls, take the Lake Moxie Road off the byway to the Moxie Falls parking area. The hike from the parking lot takes about half an hour and

Left: Sunset on Martin Pond at the Forks, Caratunk. Below: Log trailer waiting for pick-up along the Old Canada Road.

is a pleasurable walk with a very big payoff at the end. Towering pines, rushing water, and great viewing spots will leave you very satisfied. Lake Moxie is a few miles farther down the road and is a quiet lake sparsely populated with camps. It's also home to Lake Moxie Camps, one of the oldest sporting camps in Maine.

After The Forks, the byway becomes more of a wilderness area with little or no development for large stretches. This part of Maine is largely unchanged from Native American times and remote enough to be the site of a number of prisoner-of-war camps for German soldiers captured during World War II. Today this section is commonly called "Moose Alley" because you are more likely to see a moose or two standing next to the road than signs of civilization. If you do see a moose, make sure you slow down, they're a little unpredictable and will sometimes stop in the middle of the road and just look at you.

The last town on the byway before the Canadian border is Jackman. If you have the radio on you will begin to feel you are nearing the end of your trip because most of the available stations will be in French. Just south of Jackman you will see the Attean View Rest Area. This is a great place to stop and stretch your legs. On a clear day the view is vast and unobstructed, with blue sky and shimmering lakes contrasting with vast stretches of unbroken forest.

Jackman is located on Wood Pond and is a favorite of hunters and fishermen. This quiet town plays host to a loyal following of outdoorsmen who come back year after year to hunt and fish and probably to experience that elusive feeling of timelessness.

Robbins Hill Scenic Overlook, Solon.

5

Route 27

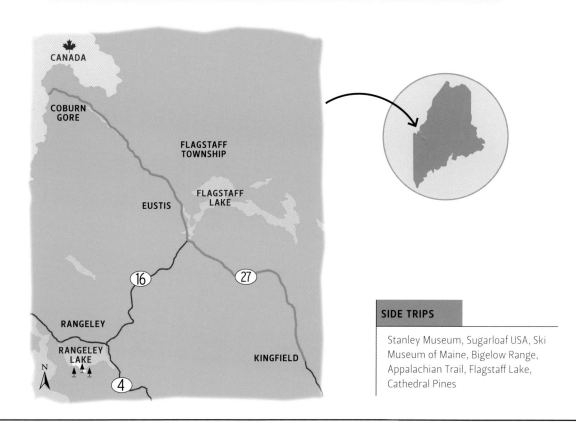

5

ROUTE 27

REGION: Lakes & Mountains
LENGTH: 47 miles
ROADS: Route 27

TRAVEL TIME: 1.5 hours
START: Kingfield
END: Coburn Gore

CANADA

COBURN
GORE

FLAGSTAFF
TOWNSHIP

FLAGSTAFF
LAKE

EUSTIS

⑯

㉗

RANGELEY

RANGELEY
LAKE

④

N

KINGFIELD

SIDE TRIPS

Stanley Museum, Sugarloaf USA, Ski
Museum of Maine, Bigelow Range,
Appalachian Trail, Flagstaff Lake,
Cathedral Pines

Previous page: Natanis Pond, part of the Chain of Ponds on the Arnold Trail. Right: Dead River Area Historical Society, Stratton. Below: Kingfield Dam on the Carrabassett River.

ROUTE 27 EXPOSES TRAVELERS TO SOME OF THE BEST SCENERY Maine's western mountains have to offer, as it winds its way through the historic Carrabassett Valley and along the Bigelow Range from Kingfield north to Coburn Gore on the Canadian border.

Forty-seven miles in length, the byway passes through Kingfield, Carrabassett, Bigelow, Stratton, and Eustis, and traverses an area rich in natural and American history. From Benedict Arnold's march to Quebec to the lost hamlets of Dead River and Flagstaff, from the forest products industry to the ski industry, from inventors and inventions to the quintessential beauty of Maine's rural farming tradition, this quiet slice of Maine is a gem well worth visiting.

Your trip begins in Kingfield, a nineteenth-century town located on a bend in the Carrabassett River. Named after William King, founder and future governor of Maine, this quiet little hamlet has a storied past that reaches far beyond its local borders. Driving into town you will notice that Kingfield is home to a Poland Spring bottling plant

— you'll see the sign but not the plant, as it is located up the hill and out of view. By necessity, industry was located along the river's edge and included numerous mills. Today the old mills are gone, but a walk across the Route 16 bridge or a stroll along the bank will give you the visual cues necessary to understand what might have been.

Kingfield is home to the Stanley Museum, located one block off Main Street in the old Stanley School. The museum celebrates the Stanley family and focuses on the most famous of the Stanleys: brothers Francis, Freelan, and their sister Chansonetta. Francis and Freelan designed and built the famous "Stanley Steamer," the first car to ascend the Mount Washington Carriage Road in New Hampshire. Another of the Stanley brothers' designs held the 1906 land speed record of 127.6 miles per hour. Chansonetta was an accomplished photographer and used a Stanley dry plate method of photography patented by her brothers and eventually sold to Eastman Kodak.

Kingfield is also home to the Herbert Grand Hotel and the Ski Museum of Maine. The Herbert, built in 1918, is a testament to the great hotels of the Victorian era. This unassuming gray and white building is located in the center of town. The Ski Museum of Maine preserves the rich history of Maine's ski industry. Ever since 1950, when Amos Winter and the "Bigelow Boys" cut the first trail on Sugarloaf Mountain and installed a rope tow three years later, the area has evolved into one of the premier skiing destinations in the country. In the last two decades the Sugarloaf complex has expanded into a year-round destination. Along with the ski complex, a nonprofit organization known as Maine Huts and Trails, headquartered in Kingfield, has established a series of backcountry trails to facilitate outdoor adventure and experiential learning. They have created a system of huts and trails and activities that allows people of all ages to enjoy the outdoors at their own pace and abilities.

Continuing north on Route 27 in Kingfield, you soon come to a small bridge spanning the Carrabassett River. You'll see

The Carrabassett River behind Kingfield Dam.

a sign that reads Public Scenic Overlook and has an arrow pointing across the bridge to the small road on the opposite side. This is the beginning of a recently completed 1.5-mile road that leads to a scenic overlook just below the summit of Ira Mountain. The road, the scenic overlook, and a strange looking open-roofed stone structure locally known as the "temple" are the creation of a man named Adrian Brochu. The road is narrow in places, unpaved, and the turn-around at the top requires some skill, but anyone willing to make the drive will be rewarded with a magnificent view of the valley. Glimpses of Route 27 are visible as it twists along the valley floor surrounded on all sides by the mountains that shadow it: Claybrook and Poplar mountains, Mount Abraham, Sugarloaf, and the mountains of the Bigelow Range.

The Carrabassett Valley was also once home to the Kingfield & Dead River Railroad. It opened in 1894 primarily to carry lumber and followed the general course of the river from Kingfield to Carrabassett Village. It closed in 1927 at the same time Maine was building Route 27, its first state highway. Today, part of the abandoned rail system has been redeveloped as the Narrow Gauge Pathway, also known as the Carrabassett River Trail. The trail runs along 6.5 miles of the Carrabassett River and is used by hikers, runners, mountain bikers, and cross-country skiers. The

Left: Carrabassett Valley seen from Ira Mountain Overlook. Right: Abandoned logging road thru the Carrabassett Valley woods. Below: Deer at Moosehorn, Coburn Gore.

trailhead can be accessed at the Bigelow Station Trailhead off Bigelow Station Road just north of the Sugarloaf Access Road on Route 27. The trail offers spectacular views of the Carrabassett River and surrounding mountains as well as many opportunities to view wildlife.

At mile 28 on the byway, the village of Stratton comes into view. Part of Eustis, Stratton is home to the Dead River Area Historical Society. The Society building is located at the intersection of Route 27 and Rangeley Road, and is dedicated to preserving the history and culture of the Dead River Area, including memorializing the lost villages of Dead River and Flagstaff. It's here at Stratton that the Arnold Trail begins, following Route 27 north to the Canadian border. In 1775 Benedict Arnold travelled up the Dead River to the Chain of Ponds on his ill-fated march to capture Quebec during the American Revolution. The trip was difficult and deadly, bad maps and inclement weather pushed the troops to near starvation and dwindled the ranks substantially.

The Arnold Expedition Historical Society has identified numerous sites along the byway and discovered an extraordinary number of artifacts associated with the expedition. While these sites aren't easily accessible to the public, persons interested in this part of American history can obtain more information and see some of the recovered artifacts at the Society headquarters in Scarborough, Maine, or on its Website. Maine state archaeologists have also discovered numerous archaeological sites along the Byway corridor that show evidence of prehistoric and historic Native American activity.

In contrast with Arnold's time, today there is a wide variety of excellent and well-marked hiking and camping opportunities available to outdoor enthusiasts. Of course, premier among the trails is the Appalachian Trail connecting some of the most popular mountains in Maine's Bigelow Range. Hikers on this section of the A.T. are presented with breathtaking 360-degree panoramic views of the surrounding area as far as Rangeley to the west and Canada to the north. Camping is also an option along the byway, from remote locations on mountain trails to sites with varying levels of amenities, such as Cathedral Pines Campground near Flagstaff Lake.

Flagstaff Lake was created in 1950 with the construction of Long Falls Dam, which controls the flow of water into the Kennebec River. The north and south branches of the Dead River meet at Eustis to form the lake. The state enforced its right of eminent domain to take over the villages of Dead River and Flagstaff, which were flooded by the lake. The inhabitants and some of the structures were relocated, but most of the buildings were destroyed. The stained glass windows and the bell of Flagstaff's church, for example, were salvaged and used in the construction of the Flagstaff Memorial Chapel located on Route 27 just outside the main village on the left.

Seventeen-thousand-acre Flagstaff Lake is now a prime fishing area with various boat-launching facilities. It is also a popular location for Nordic skiers and snowmobilers. On any given day during the winter months it's not uncommon to see the lot at Pine's Market in Eustis filled to capacity with snowmobiles waiting in line for gas and supplies.

Below: Flagstaff Memorial Chapel, Eustis. Right: Fall foliage on Flagstaff Lake.

Pine's Market is aptly named and you'll begin to notice very tall, straight red pines along both sides of the road. They are part of the old-growth stand of trees known as Cathedral Pines. More than two hundred years old and covering more than two hundred acres, these pines are one of the largest stands of old-growth forest in Maine.

After Eustis, the landscape begins to change and civilization seems to fall away. The road becomes hillier, with more twists and turns as it passes over numerous streams and wetlands, including Shadagee and Sarampus falls, emerging finally at the body of water known as Chain of Ponds. Natanis, Long, Bag, and Lower ponds are all connected by short stretches of water, creating the "chain." At Natanis Pond there is a large parking lot and overlook that gives a magnificent view of the waterway and Indian Stream Mountain, as well as the campground at Natanis Point. Leaving the overlook and continuing north, it's a short ride to Moosehorn, a small community of sporting camps, before you reach the end of the trip at Coburn Gore and the Canadian border.

Moosehead Lake

6

MOOSEHEAD LAKE

REGION: Highlands
LENGTH: 68 miles
ROADS: Routes 6 & 15

TRAVEL TIME: 3 hours
START: Greenville
END: Jackman or Kokadjo

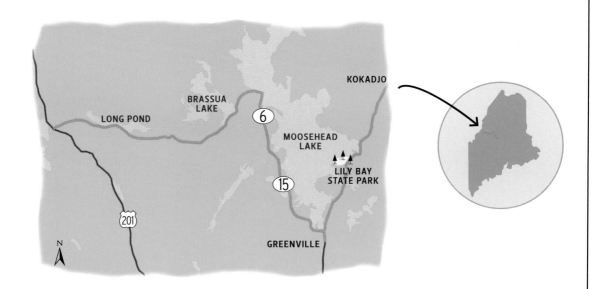

SIDE TRIPS

Moosehead Lake Marine Museum, Indian Hill Trading Post, Steamship *Katahdin*, Mount Kineo, Northern Forest Canoe Trail, Golden Road

Below: West Branch Pond Camps near Kokadjo. Right: Early morning fishing in the Piscataquis Wilderness Area.

THE MOOSEHEAD LAKE REGION IS THE GATEWAY TO MAINE'S GREAT North Woods and a destination for an increasing number of visitors. Once the land of lumberjacks, it's fast becoming a vacation spot for people seeking adventure in the outdoors. This part of Maine has traditionally been the realm of hunters and fishermen (called "sports" in Maine) who came to test their skills in the primordial forests, lakes, and rivers, staying at remote sporting camps and enjoying a rustic experience with few amenities. Today much of the region is still wild and untamed, but it offers an expanded list of outdoor activities for a wider range of people. Hunting and fishing now share the stage with sailing, hiking, skiing, snowmobiling, and other four-season activities; and today the rustic sporting camps attract photographers, bird watchers, and nature lovers, and share space with motels, hotels, and country inns.

The Moosehead Lake Scenic Byway begins at a place called Indian Hill, just south of the town of Greenville. In 1853 Henry David Thoreau visited Maine and first saw the lake from this vantage point. In his diary he wrote:

"....A suitably wild-looking sheet of water, sprinkled with small, low islands, which were covered with shaggy spruce and other wild wood, — seen over the infant port of Greenville, with mountains on each side and far in the north, and a steamer's smoke-pipe rising above a roof."

Today's view from the hill leaves much the same impression on the traveler as it did on Thoreau more than 160 years ago.

Greenville still has the feel of a small village, maybe even as Thoreau described it, an "infant port" anchoring civilization to a grand wilderness. The centerpiece of the town is the historic 1893 Shaw Block building, home to a variety of businesses, including the famous Moosehead Lake Indian Store. Greenville also has float-plane service for sightseeing or transportation into the backcountry and two outfitters, the Indian Hill Trading Post and Northwoods Outfitters, both carrying everything necessary for the adventure bound.

Located behind the Shaw Block building is the Moosehead Lake Marine Museum, and docked alongside it is another local gem and the museum's prize possession, *Katahdin*, the last remaining steamboat still plying the lake. Built in 1914, she was converted to diesel at some point and hauled logs for the timber industry until driving timber by water was discontinued in 1975. Today she still cruises the lake, but instead of logs she hauls passengers.

Greenville is also the location of the International Seaplane Fly-In and the Moosehead Lake Ice Fishing Derby. Starting with just a few planes in 1973, today the fly-in sees seaplanes from around the country showing up at the event. The fishing derby is also becoming more popular every year and is a joint effort between the Moosehead Lake Regional Chamber of Commerce and the Maine Department of Inland Fisheries and Wildlife.

Beyond Greenville, the route goes in two different directions. It continues on Route 15 along the western shore of the lake headed for Jackman and heads east to Kokadjo on Lily Bay Road.

SS *Katahdin* docked at Greenville's Moosehead Lake Marine Museum.

Following Route 15 to the west, the byway passes through a sparsely inhabited area with thickly forested terrain. Signs of logging are evident at different points and it's not uncommon to see logging trucks on the road. This section of the byway is a prime moose viewing area.

Next on the byway comes the village of Rockwood. Rockwood is located at Birch Point and for a small, out-of-the-way place, it's seen quite a bit of activity over the years, Native Americans, European settlers, wealthy sports, and the railroad have all come for one reason or another.

Birch Point lies at the narrowest section of Moosehead Lake and directly across from the giant monolith that rises out of the lake, Mount Kineo. Kineo is a volcanic plug, formed when magma became trapped and hardened in an active volcano. The volcanic cone has long since eroded,

leaving the hard flint-like rhyolite that makes up Kineo. Commonly known as Kineo flint, it was prized by Indians for use in tools and weapons, especially arrowheads. Kineo flint has been found at archeological digs up and down the East Coast, leading experts to believe tribes either made pilgrimages to the area or local Indians traded it. The Kineo massif has amazing vertical cliffs that drop more than seven hundred feet into the lake. An extraordinary site to see, it's best viewed from the steamship *Katahdin* on one of its daily tours of the lake. Another great thrill is taking the small ferry from Rockwood and climbing to the summit. The fire tower offers spectacular unbroken views of the lake and its surroundings. Climbing the tower is not for the faint of heart, but the reward is a magnificent view, so make sure you bring your camera.

The Somerset Railroad showed up in 1906, when it extended its line to Moosehead Lake, allowing sports from Boston and New York to ride directly to the new Kineo

Left: Mount Kineo seen from the dock at Rockwood. Right: Early morning on northern portion of Moosehead Lake, near Seboomook.

Station at Rockwood. The station was located in the general vicinity of today's public boat launch. From there a short steamboat ride delivered them to their final destination, the Mount Kineo House, where they could spend time in the great outdoors but still enjoy all the comforts of home.

Rockwood is an unincorporated village with no organized town government officially located in T1R1, one of Maine's unorganized territories. Strangely, though, it has its own post office and is probably the only village in the unorganized territories to have a zip code. The village also isn't named for a rock of any kind or the woods that surround the point. The story goes that a man named Rockwood placed his name on the post-office wall, and it stuck.

Today Rockwood hosts a boat launch for access to the lake, as well as a small ferry that runs to and from Mount Kineo.

From Rockwood the byway parallels the Moose River. Regulated by the dam at Brassua Lake, it's home to brook trout, lake trout, and landlocked salmon. The road hugs tightly to the river's edge along this section of the byway and has a large number of sporting camps and cottages scattered along its waterfront. Maynards, a local favorite, has been catering to sportsmen for more than ninety years and is one of the oldest sporting camps in Maine. It's also a great place to stop and have dinner. (They are open to the public.)

Left: West Branch Pond Camp (former logging camp). Right: Summit trail to the fire tower on Mount Kineo. Below: Front porch of the main lodge, Maynard's in Maine sporting camp on the Moose River.

    After Rockwood, the byway moves into dense forest-land, home to two large bodies of water, Brassua Lake and Long Pond. Brassua Lake is made of two lakes that merged when the dam was constructed in 1926. It offers good fishing and has a primitive campsite near its boat launch just off the byway. The Moose River, Brassua Lake, and Long Pond are all part of the Northern Forest Canoe Trail that connects New York, Vermont, New Hampshire, and Maine. The Maine section of the NFCT is 347 miles and runs through the Kennebec and Moose River valleys, offering a unique experience for the adventurer. This arm of the by-way ends at Jackman, located on Wood Pond. It is also the northern terminus of the Old Canada Road National Scenic

Byway (p. 43) and a favorite of hunters and fishermen who come back again and again to this region.

Backtracking to Greenville, you can begin exploring the second leg of the byway. This section follows Lily Bay Road along the eastern side of Moosehead Lake.

Leaving Greenville, Lily Bay Road rises in elevation as it skirts Blair Hill and Scammon Ridge. The landscape is mostly forested with occasional breaks of swampy areas. It passes close to Moosehead Lake at Beaver Cove, but for the most part is unbroken forest. (Look out for moose.)

Just after leaving Greenville you will see Scammon Road on the right. Scammon Road follows the southern side of Scammon Ridge and will bring anyone interested to Elephant Mountain and one of the less expected attractions of the Maine woods. On January 24, 1963, a B-52 bomber crashed into the mountain while on a training mission. The remains are still located at the crash site, as well as a memorial for the crewmembers who died in the crash.

Nine miles from Greenville, the byway arrives at Lily Bay State Park. The 925-acre park was built in 1961 on donated paper company land, giving the public access to a beautiful undeveloped section of the lake. The park is situated on a point with two protected coves and fronted by numerous small islands. It contains ninety campsites, two boat launches, sandy beaches, and walking trails. The park is also open during the winter months for snowshoeing, cross-country skiing, and snowmobiling.

Leaving Lily Bay State Park, the byway briefly hugs the lake before turning inland toward Kokadjo and First Roach Pond. The landscape along this section of the byway begins to feel increasingly remote and wild, with even fewer signs of civilization until you reach Kokadjo.

Kokadjo's welcoming sign lists the population as "not many" and seems to be fairly accurate. The village consists of a trading post/restaurant with a gas pump, a few rentable camps, and not much more. The trading post can be a very busy place, especially in the winter months when the parking lot if filled with snowmobiles. It's a very popular spot for this crowd for gassing up and a hot meal.

Another interesting side trip at the end of the byway is West Branch Pond Camps, located ten miles down Frenchtown Road from the turnoff near Kokadjo. In 1880 Charles Randle convinced the lumber company to lease him an abandoned logging camp deep in the woods on West Branch Pond and the camp has been run as a sporting camp ever since. It's possibly the oldest continuously run sporting camp in the Maine woods and is open to the public for dinner as long as you call ahead.

Kokadjo is the final stop and the end of the byway as well as the public road. It is also the beginning of the Golden Road, a private road owned by the logging companies though open to the public. The Golden Road gives sportsmen and others access to some of the most remote parts of Maine, but it has no amenities.

Left: Sunset at Lily Bay Moosehead Lake. Below: Maynard's in Maine cabins, Rockwood.

7
Katahdin Woods & Waters

7

KATAHDIN WOODS & WATERS

REGION: Highlands
LENGTH: 89 miles
ROADS: Baxter Park Rd.,
Routes 11 & 159

TRAVEL TIME: Varies
START: Baxter State Park
South Gate
END: Baxter State Park
North Gate

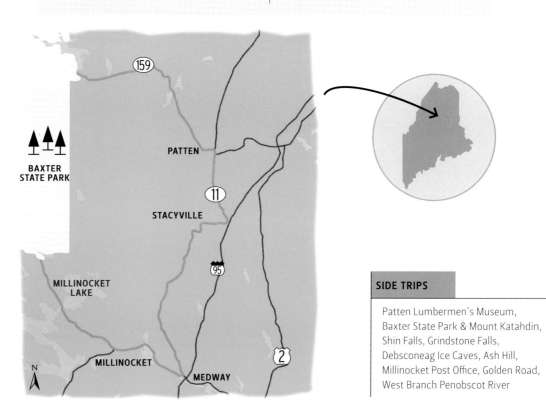

SIDE TRIPS

Patten Lumbermen's Museum,
Baxter State Park & Mount Katahdin,
Shin Falls, Grindstone Falls,
Debsconeag Ice Caves, Ash Hill,
Millinocket Post Office, Golden Road,
West Branch Penobscot River

THE KATAHDIN WOODS & WATERS SCENIC BYWAY IS EIGHTY-NINE miles long and extends from the southern gate of Baxter State Park near Togue Pond to the northern gate at Grand Lake Matagamon. The byway passes through Millinocket, East Millinocket, Medway, Staceyville, Sherman, Patten, Mount Chase, and the village of Shin Pond, exposing travelers to a variety of landscapes that includes working forests, wilderness rivers, active farmlands, historic mill towns, and panoramic vistas.

The Penobscot River Basin drains 8,600 square miles of watershed and is the second largest watershed in New England. Best known for the salmon that swim in its rivers, it offers an abundance of sportfishing opportunities. With its numerous dams and rivers, it also hosts some of the finest canoeing and whitewater rafting in the east.

Beginning at Baxter State Park's southern gate, your route heads south, passing between Upper and Lower Togue ponds on the Baxter Park Road. Lower Togue Pond is a deepwater lake with beautiful clear water and it is an excellent location for fishing. The Maine Department of Inland Fisheries and Wildlife stocks the pond with splake, a hybrid cross of male brook trout and female lake trout, creating a healthy and aggressive stock that grows to trophy size.

Leaving Togue Pond, the Baxter Park Road eventually intersects with the Golden Road at the isthmus separating Ambajejus and Millinocket lakes. The Millinocket Trading Post and Katahdin Air Floatplane Service are both located here.

Both lakes are pristine bodies of water with interesting histories. Henry David Thoreau crossed Ambajejus Lake to begin his trip up the West Branch of the Penobscot River during his 1846 trip to Katahdin; and Fredrick Edwin Church, possibly the best-known representative of the Hudson River School of landscape painting, owned a summer residence on Millinocket Lake, where he spent many years painting and fishing.

Millinocket and East Millinocket are mill towns built to supply labor and services to the paper mills that sprang up on the Penobscot River. Mills needed water for power and that was the driving force for their location; when a labor force was not located nearby, mills would build a town and import the labor. Millinocket is an Indian word that means "many islands," and actually has no relationship to paper mills.

Before leaving town you should stop at the Millinocket Post Office. Depression-Era painter John Beauchamp, working under the New Deal, painted a large mural entitled *Logging in the Maine Woods*. It's worth a visit.

The road leaves Millinocket and passes Dolby Pond, a shallow pond created by the construction of Dolby Dam, built to generate power for the East Millinocket mill. The town of East Millinocket is designed as a rectangle — all roads run north/south or east/west — laid out and built by the mill owners to house the labor force required to operate the mill.

Beyond East Millinocket the byway continues on Route 11 to the town of Medway, which sits at the confluence of the east and west branches of the Penobscot River. Every August, Medway hosts the Wooden Canoe Festival, paying homage to the art of building wooden canoes.

Katahdin seen from Sandy Stream Pond, Baxter State Park.

After leaving Medway and crossing the bridge that spans the East Branch the route turns left up Grindstone Road (following Route 11). It follows the river through a thickly wooded area showing signs of past logging activity. The road sits tightly to the river until it reaches a small cluster of houses and buildings known as Hay Brook. Then it veers off to the northeast and continues through dense forest until reaching the Grindstone-Staceyville area, part of the Herseytown Unorganized Territory. Considerable acreage has been cleared here and numerous farms are located on both sides of the road. Some fields seem inactive and are filled with wild vegetation in summer, especially goldenrod.

Route 11 follows a fairly straight line through this area, until it meets the old Matagamon Tote Road, where it takes an abrupt right and heads east, continuing on a while before taking a left and heading north again. The route follows this pattern once more as it passes through Stacyville, an active farming and logging district, before reaching Station Road in Sherman, headed for Patten.

On Station Road the byway soon enters Sherman Station, a small village that was once part of Maine's largest commercial enterprise — lumbering. The Bangor and Aroostook Railroad passes through the center of this working village, though, like the industry, the village feels as though it's slowed in recent years. Large railroad buildings, now idle, sit next to a siding just off the main track, and on the opposite corner of the road sits a now quiet lumber mill. From here the road climbs Ash Hill and runs along the ridgeline, affording spectacular views of the surrounding countryside. From the top of the hill you're rewarded with a 360-degree vista of the surrounding countryside —Katahdin to the west, the town of Patten to the north, and picturesque farmland to the south and east. The Thousand Acre Bog is located to the east but hidden behind the forested slope that abuts acres of farm fields. The six-mile-long Sherman Patten Multi-use Trail follows an old railroad grade through a fairly remote section of these

Hay Bales along the South Patten Road (Route 11).

woods, passing close to the western side of the Thousand Acre Bog and is a prime location for viewing wildlife.

Located at the bottom of the north side of Ash Hill is the town of Patten. Another stop on the B & A Railroad, Patten, like Sherman Station, is a mill town. And like many other towns in northern Maine, Patten has an idle mill and a few vacant lots exhibiting rusting logging equipment, vestiges of a different time. However, the town also has a gem, a special place that honors the men and machines that helped build an empire – The Patten Lumbermen's Museum.

The museum is located on Shin Pond Road and its mission is to preserve the memories and methods of Maine's logging heritage. It consists of nine buildings that house a large collection of tools and machines used in the logging industry in pre-World War II Maine. Two of the buildings are full-scale reproductions of logging camps. One building is circa 1820 and closely resembles the camp that Henry David Thoreau described in his 1846 trip to Maine:

These camps were about twenty feet long by fifteen wide, built of logs, – hemlock, cedar, spruce, or yellow birch, – one kind alone, or all together, with the bark on; two or three large ones first, one directly above another, and notched together at the ends, to the height of three or four feet, then of smaller logs resting upon transverse ones at the ends, each of the last successively shorter than the other, to form the roof. The chimney was an oblong square hole in the middle, three or four feet in diameter, with a fence of logs as high as the ridge. The interstices were filled with moss, and the roof was shingled with long and handsome splints of cedar, or spruce, or pine, rifted with a sledge and cleaver.

The museum's other camp building is an authentically furnished double camp building split between sleeping quarters and dining room. The museum also features a saw mill and outbuilding filled with logging tools and equipment, including a batteau, the log driver's standard working boat.

Beyond Patten the route follows Shin Pond Road, traveling through Mount Chase Plantation toward Upper and Lower Shin ponds. The two ponds are connected by a small

Right: Winter pines along Shin Pond Road. Below: Carved sculpture at the Lumbermen's Museum.

stream and both are good fishing spots for trout and land-locked salmon. Shin Pond Village, a family-run camp-ground, sits just on the other side of the bridge on Lower Shin Pond. It offers hunting, fishing, canoeing, kayaking, and wildlife viewing.

From the campground the road continues through dense forest until it reaches Grand Lake Road, which continues on toward Baxter State Park. The road eventually crosses the Seboeis River and there's a picnic area on the right just after the bridge. From here the road continues through thick forest that eventually becomes overshadowed by the huge cliffs of Dark Horse Mountain. Soon wild and undeveloped Grand Lake Matagamon, sitting at the base of the mountain's northern slope, comes into view and this trip ends at the north gate to Baxter State Park.

Acadia All–American Road

8

ACADIA ALL-AMERICAN ROAD

REGION: Down East & Acadia
LENGTH: 40 miles
ROADS: Route 3, the Park Loop Road

TRAVEL TIME: 3 hours
START: Trenton
END: Bar Harbor

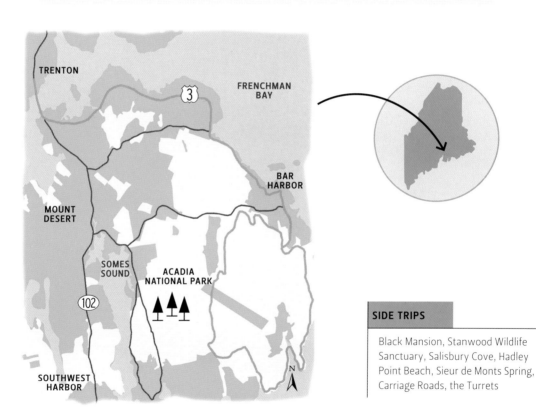

SIDE TRIPS

Black Mansion, Stanwood Wildlife Sanctuary, Salisbury Cove, Hadley Point Beach, Sieur de Monts Spring, Carriage Roads, the Turrets

Previous page: Bar Harbor sunset. Above: The schooner *Margaret Todd* docked in Bar Harbor.

THE ACADIA ALL-AMERICAN ROAD BEGINS AT THE TRENTON- Ellsworth line and at six miles passes the Thompson Island Information Center, run by the National Park Service. The center is located on the left-hand side of the road on Thompson Island just before driving onto Mount Desert Island. Ellsworth is the commercial center for the region and includes restaurants, retail shops, adventure outfitters, and a marina, but is also home to unique historic landmarks. The 300-acre Black Mansion, known sometimes as Woodlawn, is located on Surry Road and is open to the public. Colonel John Black amassed a fortune in the timber industry and his home has been preserved as an example of the wealth enjoyed by some in the 1800s. The mansion is furnished with authentic period furniture and offers guided tours.

View toward Champlain Mountain from Sand Beach.

Another interesting place is the 200-acre park known as Birdsacre, a wildlife center officially named the Stanwood Wildlife Santuary. Located a quarter mile east of the Route 1–Route 3 intersection, the park includes the homestead and gardens of Maine ornithologist and photographer Cordelia Stanwood. It's also home to Woodland Gardens and the Richmond Nature Center, educational facilities that help preserve Maine's natural heritage. This initial section of the byway also offers occasional views of protected inlets and the larger bay, which hint at the adventure that lies ahead.

If you decide not to stop at the Thompson Island Visitors Center, you may opt to visit the Acadia National Park main visitors center at Hulls Cove, where you'll find lots of pamphlets as well as a short movie about Mount Desert Island and the park.

From here the byway continues along Route 3 until it enters Bar Harbor, the most popular town on the island. Its main street is populated with locally owned stores and restaurants, and a short walk to Grant Park and the pier reveals activities such as whale watching and sailing. There are also hotels, motels, and roadside cabins available for those who plan to stay for a while. Two places at Bar Harbor that are often overlooked are Salisbury Cove and Hadley Point Beach. Salisbury Cove is located off Old Bar Harbor Road. This quiet little spot is a perfect place to relax and enjoy the island, especially when Route 3 is busy.

Hadley Point Beach is located at the end of Hadley Point Road. It's a small triangular beach that offers swimming opportunities and is a popular place for kayaking. Coastal Kayaking Tours, located on Cottage Street in downtown Bar Harbor, runs tours from this beach.

THE PARK LOOP ROAD

Much of the Park Loop Road was designed by Frederick Law Olmsted Jr., son of the famed architect who designed

Boston's Emerald Necklace and New York's Central Park. Acadia National Park takes up about 35,000 acres on Mount Desert Island and if you don't have an extended amount of time to see the sights, the Park Loop Road is your best bet.

The road is twenty-seven miles long and has a maximum speed limit of thirty-five miles per hour. There are numerous parking areas along the roadway and because most of the road is a one-way, two-lane road, travelers are allowed to park in the right-hand lane along most of the loop.

THE PRECIPICE

The Precipice Trail is located on the east side of Champlain Mountain, about a mile north of the Schooner Head entrance and when it's not closed to protect nesting peregrine falcons, it is by far the most challenging climb in the park. The Precipice is a 1,000-foot climb up exposed vertical cliffs on the ocean side of Champlain Mountain. In places the climber is required to use short bridges and iron ladder rails that have been drilled into the pink granite. This climb is not for the faint of heart or for kids and shouldn't be attempted in inclement weather.

Left: Carriage Road bridge crossing the Park Loop Road. Right: Sand Beach in Winter. Below: The author's children looking toward Sand Beach from the Beehive.

THE BEEHIVE

This large chunk of pink granite sits across the road from the Sand Beach parking lot. At 520 feet, it isn't the highest or longest climb in the park, but it's probably one of the more skilled — code for dangerous — climbs in the park.

The trail begins gradually but rises abruptly when it turns at the wooden trail marker in the woods. There should be a sign at this point that reads "slippery when wet," because, like the Precipice Trail, it shouldn't be attempted in inclement weather. The trail requires the use of iron ladder rungs similar to those found on the Precipice Trail, and iron grates have been put in place to fill in gaps in the ledge.

The view is spectacular from the summit and it's really impressive when you look down at Sand Beach and see the people, but this climb isn't recommended for kids, or adults who have a fear of heights. If you feel you or your kids must see the view from the summit but can't bring yourself to scale the eastern face, there is a must less vertically challenging trail that leads around the back of the Beehive.

Jordon Pond with the Bubbles in the background.

SAND BEACH AND GREAT HEAD

Sand Beach sits across the road from the Beehive and is tucked between rocky ledges. The term sand is a misnomer in this case, because the "sand" on this beach is 85 percent seashell fragments and only 15 percent granite — what we usually think of when we say sand.

It's not uncommon even in the height of summer to see a lot of people exploring the beach but very few actually in the water. That's because the temperature of the ocean here rarely gets above fifty-eight degrees. Located at the far end of the beach is the Great Head Trail. The trail loops around the outer peninsula and offers some extraordinary views of the ocean, Sand Beach, the Beehive, and beyond.

THUNDER HOLE

Next on the Park Loop is Thunder Hole. If you come at the right time of day, with the right wave action and wind, you're in for a treat. Over time waves have carved a small inlet shaped so that when water rushes in, air and watery spray are violently released, making a loud clap. The Park Service has installed safety railings near Thunder Hole, but the rest of the ledges are open for exploration.

A short walk south from Thunder Hole is Monument Beach. There are no trails leading to the beach, so it is difficult to access, but views from above are impressive. The water has carved out stone pinnacles on either side of the cove and has pounded the rocks that make up the small beach into smooth round stones called cobbles. The stone monuments remind you of something you might see in a Western movie.

OTTER CLIFF

At 110 feet high, Otter Cliff is the highest coastal headland in North America. The cliff is less than a mile from Thunder Hole and a favorite of rock climbers. Like Monument Beach, the area around the bottom of the cliff is strewn with cobbles rounded by wave action.

Submerged in front of the cliff is a rock formation called the Spindle, which is a hazard to navigation. Samuel de Champlain ran aground on the Spindle in 1604 during his exploration of the coast and had to spend the winter in Otter Cove to repair his ship.

LITTLE HUNTERS BEACH

Little Hunters Beach is one of the best-kept secrets of the Park Loop Road. There are no signs to indicate its location. Past Otter Cliff, around Otter Cove, over one of Olmsted's magnificent bridges, and beyond the Blackwoods Campground area is a small bridge over a runoff stream. Just before the bridge is a small pull-off that can accommodate about three cars. Park at the bridge and head down the stairway on the ocean side of the road and you will be on Little Hunters Beach. The beach is made up of millions of small cobbles of different shapes and colors. When waves break on the beach, the surf makes a unique sound, creating a calming effect on anyone who sits awhile. The pebbles are so interesting, the National Park Service has placed a sign telling people not to take the stones.

WILDWOOD STABLES

Wildwood Stables is located half way between Little Hunters Beach and Jordan Pond and offers a variety of carriage rides and tours. The carriage roads were designed and built by John D. Rockefeller, Jr., and located on his property. A variety of carriage tours are available, including the Tea and Popover Carriage Ride. The horse-drawn carriage takes you to Jordan Pond, where you enjoy tea and the Jordan Pond House's famous popovers

Left: Lupine flourishes along Maine's roadsides. Right: Monument Beach. Below: Magic stones on Little Hunters Beach — they sing when the waves come in.

while enjoying vistas of the pond and the famed Bubbles from a table on the back lawn.

THE JORDAN POND HOUSE

The Jordan Pond House is located at the southern end of Jordan Pond. Slightly elevated, its famous lawn has a perfect view of the water and surrounding mountains. The Jordan Pond House is a full-service restaurant serving lunch, dinner, and afternoon tea. It is the only restaurant of its kind in the park and is open from mid-May to late October.

There are pathways through the woods and trails along the pond's edge. The water is crystal clear and deep and it's a good fishing pond for lake trout and landlocked salmon. Small boats are allowed.

CADILLAC MOUNTAIN ROAD

Beyond Jordan Pond the Park Loop Road passes Bubble Pond and then runs along the ridge paralleling Eagle Lake on its way back to its starting point. It also passes the entrance to Cadillac Mountain Road. The road to the summit of Cadillac is not part of the Park Loop Road, but we can't talk about this byway without including Cadillac Mountain.

With a summit of 1,530 feet, Cadillac is the highest mountain in the park and the highest on the eastern seaboard. The road snakes up the mountain in a series of traverses and intermittent sharp hairpins until it reaches the parking lot at the summit. The mountain is composed of the pink granite common in the park and covered with windblown spruce and pine stunted by the harsh winds. On a clear day the view from the top is unobstructed and awe inspiring. The Porcupine Islands, Bar Harbor, Frenchman Bay, the Cranberry Isles, and the vast Atlantic Ocean are laid before the traveler standing on the summit. The mountain road is open twenty-four hours a day, but only from April 15 through November 30. It's a great place to stargaze and, while not officially part of the byway, it may be the best spot to end your trip.

View of Frenchman Bay and the Porcupine Islands from the summit of Cadillac Mountain.

Schoodic National Scenic Byway

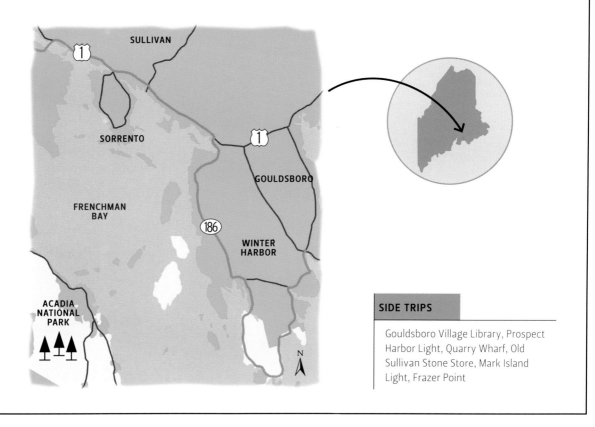

9

SCHOODIC NATIONAL SCENIC BYWAY

REGION: Down East & Acadia
LENGTH: 29 miles
ROADS: Routes 1 & 186

TRAVEL TIME: 1 hour
START: Hancock
END: Prospect Harbor

SULLIVAN

1

SORRENTO

1

GOULDSBORO

FRENCHMAN
BAY

186

WINTER
HARBOR

ACADIA
NATIONAL
PARK

N

SIDE TRIPS

Gouldsboro Village Library, Prospect
Harbor Light, Quarry Wharf, Old
Sullivan Stone Store, Mark Island
Light, Frazer Point

Previous page: Pink granite along the Acadia National Park section of the Schoodic Peninsula. Above: Loading traps. Below: Harbor seals on Frenchman Bay.

A SHORT RED LINE PASSES THROUGH TWENTY-NINE MILES OF THE Schoodic Peninsula on the official road map of Maine and although not a great distance as "Maine miles" go, this little line highlights some of the most profoundly striking countryside in America. This is the Schoodic National Scenic Byway, one of only four nationally recognized byways in Maine. Sloping toward the ocean and composed of pines and firs, pink granite, wildflowers, and a deep blue sea, this special piece of land harkens back to a time when the clock ticked slower and life moved a little easier. This is nature without the props and schedules, without the thick informational flyers and commercial tourist enterprises. When you visit this part of Maine, you will learn how to exhale again — the vista demands it.

The byway begins at the town of Hancock on the western shore of the Taunton River, a bio-diverse tidal river rich in marine life and home to a wide array of plant species and

Lobster traps waiting for spring.

marine organisms. Pick up a map at the information/rest area, the official western gateway to the byway, located on your left just before going over the Taunton River Bridge to begin your journey.

Once over the bridge you've entered Sullivan, Maine. On this part of the journey the road travels through spruce and fir wooded sections of Route 1 with frequent views of the upper reaches of Frenchman Bay. You can now shut the air conditioner off, open the windows, and slow down. Stick your hand out and let the wind blow through your fingers, smell the salt air, this is Down East Maine — more rural, less crowded, and quieter. Look around and you will begin to see evidence of old farm fields, wood lots, historic homes, old boats, and signs of lobstering.

You will soon come upon Route 186 on your right, also known as South Gouldsboro Road. As you take the turn, the road might feel a little confined because of the hedges and trees close to the roadway, but the route opens up after you cross the small bridge over the inlet connecting Jones Pond to Frenchman Bay. The pond is located in Gouldsboro and is open to the public, with access on its eastern shore. With the exception of the town of Winter Harbor, Gouldsboro covers the entire Schoodic Peninsula and then some. It includes the villages of South Gouldsboro, West Gouldsboro, Prospect Harbor, Birch Harbor, and Corea.

Take a slight detour and drive around to the other side of Jones Pond and look for the aptly named Recreation Road; at the end of this road you will find a picnic area with tables, a boat ramp, and swimming area.

The byway continues on past the Jones Pond area with more views of forest and fields until it reaches Winter Harbor, a former village of Gouldsboro that incorporated as a town in 1895. It's home to a small but proud fishing fleet and picturesque Mark Island Light, better known as Winter Harbor Light. The lighthouse was built in 1854 and sits proudly on its small chunk of coastal granite with ex-

pansive views of Mount Desert Island and Acadia National Park as a backdrop.

At this point, the byway turns closer to the shore as it enters the Schoodic section of Acadia National Park, the only section of the park located on the mainland. Here the park has a six-mile one-way loop road with lots of pull-outs for you to stop and explore. The first place of interest to many is Frazer Point. There's a picnic area with tables, grills, and a dock, but even if you don't plan to picnic or fish this is a great place to explore. Much of the pink granite in this area is laced with strips of black basalt and makes for interesting viewing. There are also numerous wild flowers that grow in this salty, sun-baked terrain. *Rosa rugosa*, commonly known as beach rose, can be found growing at the sunny edges of the scrub pines that hold the last vestiges of soil to the edge of the rocky shore. You'll also spot slender blue iris growing in the green zone that bumps up against the sea. These, along with many other wildflowers tolerant of this salty climate, are there to be discovered for those who decide to look.

There are many other spots to pull over and check out the view, including Schoodic Point and Blueberry Hill.

Left: Mark Island Light, also known as Winter Harbor Light. Right: One of numerous smalls islands scattered around the bay. Below: Beach roses growing along the byway.

Schoodic Point is probably the biggest attraction on the byway. It's a destination for many people who come to view the impressive vistas or to just sit and meditate. It's an amazing place with granite ledges, outcrops, caves, and cliffs, all impressive and wonderful to explore. Tide pools abound at low tide and are filled with various strange looking sea plants and creatures. It's a great place for kids to explore but be careful and keep an eye on them, waves are big enough at times to wash the unsuspecting visitor right off the rocks and out to sea.

Blueberry Hill, just a little way down the road from weather-beaten Schoodic Point, will give the visitor a very different perspective of the shoreline. The byway turns up the eastern coast of the peninsula and gives the visitor a more protected view of the ocean and its surroundings. From this elevated position you can look down at a beach covered with smooth stones of all sizes and an ocean of varying shades of blue.

Birch Harbor is at the edge of the park, but the byway continues on to Prospect Harbor. This quiet village is home to a small fishing fleet and Prospect Harbor Light. The lighthouse is located on a short spit of land owned by the Navy and is best viewed from across the way on Route 186.

Prospect Harbor was also home to the last sardine cannery in the United States, known as the Stinson Seafood Plant, it closed in 2010. More than four hundred canneries resided up and down the Maine coast, they were a hallmark of coastal community life and an economic engine for many communties. It was once a common site to see women and children exiting their homes and walking down to the docks as the factory horn blew, some carrying their own scissors, to take up their working stations in the processing plants. In 2011, a year after closing, the Stinson plant reopened as a live lobster company, only to close again in 2012 when the new owners couldn't make a go of it.

Old house with wharf in Gouldsboro.

Black Woods Scenic Byway

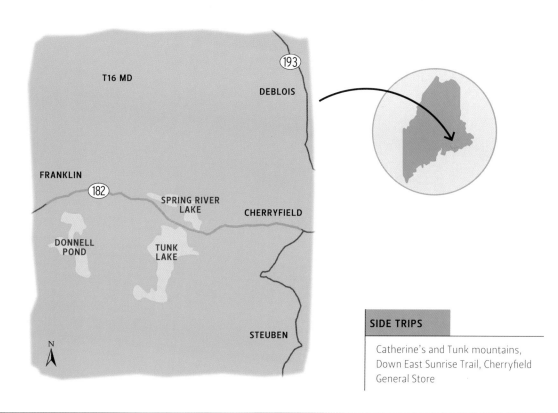

10

BLACK WOODS SCENIC BYWAY

REGION: Down East & Acadia
LENGTH: 12.5 miles
ROADS: Route 182

TRAVEL TIME: 0.5 hour
START: Franklin
END: Cherryfield

T16 MD

193

DEBLOIS

FRANKLIN

182

SPRING RIVER LAKE

CHERRYFIELD

DONNELL POND

TUNK LAKE

STEUBEN

N

SIDE TRIPS

Catherine's and Tunk mountains, Down East Sunrise Trail, Cherryfield General Store

Previous page: October mist on Tunk Lake. Right: View of Hog Bay from Route 182. Below: Trees overhanging Black Woods Road (Route 182).

THE 12.5-MILE-LONG BLACK WOODS SCENIC BYWAY BEGINS AT THE border between Franklin and the Unorganized Territory of Township 9. Located in the heart of the Hog Bay and Taunton Bay watershed, this route traverses a rich and vital ecosystem important to the region and to the people who live and work there. The byway begins at a fork in the road, in a small hamlet at the divergence of Routes 182 and 200. Hog Bay Road branches to the right and brings you back to Route 1; Black Woods Road forks to the left and brings you through primitive landscape with spectacular views, eventually ending in the historic town of Cherryfield — blueberry capital of the world. As Robert Frost wrote, "Two roads diverged in a wood, and I — I took the one less traveled by, and that has made all the difference." Take his advice and go left at the fork in the road.

Some people believe the byway was named for the trees that blanket the roadside, trees that seem at times to canopy the road with their dense foliage. It's more likely, however, that the area was named after Colonel John Black, a man who managed timber rights in the region for eastern lum-

Blueberry barrens on Georges Pond Road.

ber barons and amassed a small fortune of his own in the process. Black Woods Road probably started out as Black's Woods.

Today, you are more likely to come upon cultivated blueberry barrens than timber harvesting. Lumbering has slowed significantly in this part of Maine, and, true to form, Mainers discovered another sustainable product created by Mother Nature and ripe for the picking — wild blueberries.

Shortly after starting down the byway you will notice George's Pond Road on the left. Turning left and motoring up this short hilly road brings you right into the middle of a large barren of wild blueberries.

There are more than 60,000 acres of blueberry barrens in Maine and this is a prime example of what they look like. Wild blueberries grow in the understory of the forest and compete with all the other plants that grow in Maine's rocky, acidic soil. To create a productive field for commercial harvesting, the industrious farmer must first log off the trees, remove the stumps, kill the remaining non-blueberry undergrowth all with minimal disturbance to the existing blueberry plants. It can take up to ten years to create a productive field for profitable harvesting and although the blueberry plants are considered wild, their fields require care. Blueberries are managed on a two-year cycle and each fall after the harvest the producing fields are either mowed or burned to kill pests, remove detritus, and maintain healthy plants. The alternate year for the plants is a growth year. Low-bush blueberry plants initially grow from seed, but once established begin to send out runners that develop root systems, spreading the plant over an extended area of the barren. This lateral growth from one plant can take up as much as 250 square feet of surface area.

You may feel rewarded by the view from the edge of the barren, but it's well worth the short drive or walk up to the top of the hill. Spread out in panorama are the mountains that surround the byway: Schoodic, Caribou, Black, and Tunk mountains can all be seen from this wonderful vantage point.

Once back on the drive, the road twists and turns through alternating dense forest, wetlands, bogs, and marsh. The road also passes numerous lakes and ponds created by the retreating glacier that covered New England during the last ice age. The combination of these wetlands and waterways helps filter runoff, creating some of the purest water in Maine, and ponds and lakes in the area, including Tunk Lake, were the site of early ice harvesting operations. Once cut, the large blocks of ice were stored in ice houses and insulated with sawdust against the summer heat.

The first significant body of water you will happen upon is Fox Pond, a popular spot for swimming and fishing. Tucked up tight to the water's edge, the road follows the pond's curves, creating a visually pleasing experience. This is a good place to stop and take a break. There's a gravel turnout at a curve in the road just before the byway dips, and another larger one at the northern end of the pond. If you're interested in bird watching this is a good spot to sit for a spell; you might spot one of the loons that nests here or an eagle circling the nearby barrens looking for lunch. After leaving Fox Pond, the road begins a steady climb and traverses the northern slope of Catherine's Mountain. Tunk Mountain is on your left and Catherine's Mountain on your right. Tunk Mountain is treeless and offers a fine view of the surrounding landscape, including the Atlantic Ocean to the east and, on a clear day, Katahdin to the west. Anyone deciding to climb Catherine's Mountain may also discover evidence of old pits, quarries, and mines; the remnants of nineteenth-century mining operations in search of gold silver and molybdenum. The road along this section is also said to be haunted. Legend has it that a headless woman wanders the road at night and anyone who sees her is required to stop and offer her a ride or suffer the consequences.

At the halfway point you reach Tunk Lake. There is pub-

Right: Blueberry barrens seen from the air.
Below: Fox Pond.

lic access at a small parking area that includes a boat ramp and picnic tables. To the left of the boat ramp along a short trail is a small stream that drains crystal-clear water into Spring River Lake, a testament to the purity of local waters.

Farther down, the road crosses over Tunk Stream. Some sections of this stream are canoeable at high water, but bring your insect repellent and be prepared to portage some sections. Spring River Lake, Long Pond, and Round Pond are located along this section and are less accessible from the byway, though Long Pond has a boat launch with

parking. There are also a large number of hiking trails that crisscross the area.

The Black Woods Scenic Byway connects in two places to the Down East Sunrise Trail, an 85-mile-long former rail corridor connecting Brewer to Calais. The gravel bed is all that remains of the track and the trail is open to all ages and ranges of activity, from horseback riding to bicycling to walking. It gives you a new perspective of the rural landscape.

Once you pass over the eastern intersection of the rail trail you enter the town of Cherryfield. The Narraguagus River bisects the town and was a great aid to early industry. Originally known for lumbering and farming, the town was an oasis of industry in the middle of a primitive landscape. With some fifty-five historic properties, most located in the downtown area, today Cherryfield is one of the best examples of a nineteenth-century Maine village. Many of the names associated with these historic structures have the familiar ring of old New England families: Patten, Burnham, Coffin, Adams, but the most prominent name by far is Campbell.

General Alexander Campbell was granted three hundred acres of land for his service in the French and Indian War, moved his family to Maine (then still part of Massachusetts), and they've lived here ever since.

One of the most interesting buildings in town is the Cherryfield General Store. It was erected in 1865 and originally operated as a general store. Over the years it has been host to a diverse number of businesses, but today it's back to being a general store; reclaimed and resurrected by its current owner and selling local goods and products.

The two towns that anchor the Black Woods Scenic Byway are not officially part of the 12.5-mile-drive, but they play a vital role in understanding and appreciating the intrinsic beauty and history of this route. They help travelers define their experience and that is an essential part of what makes this roadway great.

Left: The historic Frank W. Patten Store, now the Cherryfield General Store. Below: Looking across to the historic A. Campbell & Associates building, now the Narraguagus Trading & Antiques Shop.

The Million-Dollar View

11

THE MILLION-DOLLAR VIEW

REGION: Down East & Acadia
and Aroostook
LENGTH: 8 miles
ROADS: Route 1

TRAVEL TIME: 1 hour
START: Danforth
END: Orient

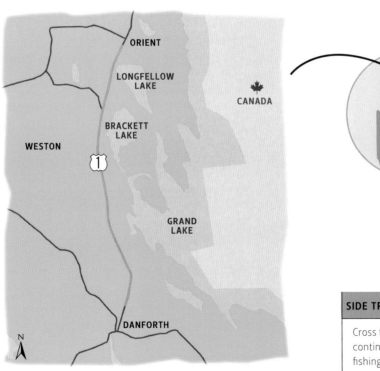

SIDE TRIPS

Cross the border into Canada or
continue north into Aroostook County,
fishing and hunting excursions

A WARM WELCOME AWAITS TRAVELERS IN THE RURAL communities along this stretch of U.S. Route 1. The eight-mile-long Million-Dollar View Scenic Byway is one of the oldest in Maine. Established in 1971, this short stretch of road didn't require a corridor plan until 2003, when the byways program became more formalized.

Danforth in the south and Orient in the north anchor the byway, but together only make up about three quarters of a mile of the designated byway. Nearly the entire course is located in Weston. In area, Danforth is one of the largest towns in Maine, covering more than sixty square miles. It's population, however, is less than six hundred. Once a prosperous lumbering community, ushered in on the shoulders of the railroad, today Danforth's economy is based mostly on outdoor tourism such as fishing, hunting, and snowmobiling, and there are several campgrounds and cabins for these activities. The town also hosts an annual summerfest.

With a population of less than three hundred, Weston is even smaller than Danforth. It, too, is primarily a tourist town, and the beauty of the byway is the main draw — its Website states, "Welcome to Scenic Weston."

Your short drive ends in Orient. With fewer than two hundred year-round residents, it's even smaller than Weston. Orient, however, is home to East Grand, North, Deering, and Longley lakes, making it a fishing mecca, and the population swells seasonally as people come to enjoy the area.

The name of this byway really does say it all. The view from the top of Peekaboo Mountain — more a ridge than a mountaintop — is spectacular. There are two turnouts with parking and information panels. The view west from the first turnout is unimpeded and Katahdin is clearly visible jutting into the sky, even though it's a great distance from this part of Maine. The second turnout, located on the right side of the road, has clear views of the Chiputneticook Lakes, including East Grand, Brackett, and Deering. The fishing in these waters is as good as it gets anyplace in Maine and is a destination point for anglers. The views looking east are even more impressive than to the west and you can clearly see Canada just across the way. You may be lucky enough to spot wildlife, including bears, deer, eagles, loons, and moose.

Potato barn and corn fields near Danforth.

The Bold Coast

12

THE BOLD COAST

REGION: Washington County
LENGTH: 125 miles
ROADS: Routes 1, 187, 189, 190 & 191

TRAVEL TIME: 6 hours
START: Milbridge
END: Eastport

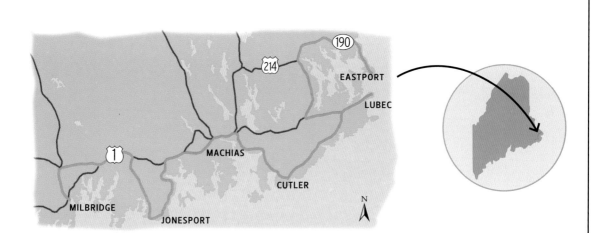

SIDE TRIPS

Down East Sunrise Trail, Ruggles House, Ice Age Trail, Wild Blueberry Land, Moosehorn Wildlfe Refuge

Previous page: Lobsterboats in Jonesport.

ALTHOUGH PEOPLE DEBATE WHERE DOWN EAST BEGINS AND the rest of Maine leaves off, it's safe to say if you're in Washington County, you are officially Down East. The Bold Coast Scenic Byway is 125 miles long and is anchored by Milbridge in the south and by Eastport and Lubec in the north. Milbridge was a hub of commerce, and boatyards were located all along the Narraguagus River, a tidal river situated on the town's eastern shore. Timber was big business in Maine during the nineteenth and early twentieth centuries and much of it was managed and regulated in shipping towns like Milbridge. If you have time, a point of interest worth visiting is McClellan Park. This oceanside park offers spectacular views of Pleasant Bay and has the same look and feel of more visited places like the Schoodic Peninsula and Acadia National Park.

From Milbridge the byway moves up Route 1, following the eastern shore of the Narraguagus River until reaching the town of Cherryfield. The Bold Coast and Black Woods scenic byways connect at Cherryfield, a well-preserved nineteenth-century village with more than fifty-five properties on the historic register. Modernization passed right by Cherryfield and preserved an important piece of Maine's heritage for future generations. Many of the buildings have been saved because generations of families have occupied the same property and long understood their intrinsic value.

Cherryfield is a good place to stop and walk around. A map of the historic district can be obtained from the Cherryfield-Narraguagus Historical Society, located at 88 River Road, or online at cherryfieldhistorical.com/walking/cnhs-tour.shtml. The small commercial town center has some unique buildings, including one of the original general stores, still open for business and selling local products. It is the farthest thing from a souvenir shop you'll find in Maine.

The Sunrise Trail (sunrisetrail.org) connects to the byway at Cherryfield. The trail is an 85-mile off-road multi-use trail that runs along the entire Down East coastal area and is open to the public for recreation.

Continuing on, the byway enters the town of Harrington. Not unlike many other coastal towns, Harrington is a quiet little fishing village located on a tidal river known mostly for its maritime heritage. Its harbor is home to lobstermen and clam diggers and is a tranquil place to spend the day fishing from the shore or off the town pier. Harrington is also a good place to try your luck at salmon fishing on one of the local rivers — the season starts in early May and ends in early June. A nice spot to stay for a day or two is Sunset Point, which offers a choice of campsite or quintessential oceanside cabin.

Leaving Harrington, the byway passes over a large plateau that is the center of the blueberry industry in Maine. A side trip is recommended for anyone wanting to better comprehend the vastness of the blueberry barrens and

McCurdy's Smokehouse in Lubec, the last canning factory building in Maine.

their economic influence on the region. A ten-mile loop drive on Station/Epping Road will afford expansive views of boldly colored barrens carpeting the landscape as far as the eye can see; green in spring, blue in summer, and red in winter. Look for it on the left a few miles before Columbia Falls. The road actually forms a horseshoe — if you stay on Epping Road, you'll end up back on Route 1 a little farther down from where you first turned.

Columbia Falls is home to the historic Ruggles House (ruggleshouse.org). Thomas Ruggles was a wealthy lumber dealer, postmaster, captain of the militia, and court judge, and construction began on this elaborate Federal-style home in 1818. It is a good example of the contrasting wealth present during the 1800s between the barons of industry and the majority of the local inhabitants.

Columbia Falls is also home to the somewhat more eclectic Wild Blueberry Land. The main structure is a large blue dome known as the world's largest blueberry. Its collection of strange accessories, all painted blueberry blue, pay homage to the wild blueberry. Everything in the store, including the coffee and the ice cream, have a blueberry theme. Kids love this place, so if you have the time, stop, have a blueberry ice cream, play a round of blueberry miniature golf, and buy some blueberry flavored coffee for the ride.

The byway leaves Route 1 at Wild Blueberry Land and turns down Route 187 to Jonesport, an active fishing village with a long history of making a living from the sea. It was once home to a number of sardine canneries, but they are gone now and the major business is lobstering. The eponymous lobsterboat design seen up and down the New England coast is based on the Jonesport-Beals Hull, which

Left: Wild Blueberry Land, Columbia Falls.
Right: Blueberry barrens, Columbia Falls.

originated here. To really appreciate the town you should cross the Beals Island Bridge. Looking back from the bridge you get a real feel for the town's layout, as well as some fantastic views of the village and harbor.

After Jonesport, continue on Route 187, which offers stunning views of Englishman Bay and beyond to Roque and Spruce islands. This section of the trip is filled with fantastic views of blueberry barrens combined with an abundance of roadside wildflowers and ocean vistas.

Shortly the byway returns to Route 1 and enters the village of Jonesboro. Roque Bluffs State Park is just south of the town center and is worth a visit. The 270-acre park is located on Schoppee Point along Englishman Bay, a quiet out-of-the-way place that is a favorite of bird watchers. The unusual geologic formations of the park make it an official stop on Maine's Ice Age Trail (iceagetrail.umaine.edu).

From Jonesboro continue on Route 1 to Machias, where the first naval engagement of the Revolutionary War took place. The historic Burnham Tavern is located on a hill overlooking the river and it was here that the plot to capture the British warship *Margaretta* was planned. After a short but furious engagement colonials captured the British vessel, leading the battle to be dubbed "The Lexington of the Seas."

The Bold Coast Byway is rejoined by the Sunrise Trail in East Machias. East Machias has three large freshwater lakes: Gardner, Second, and Hadley, which are largely undeveloped and are home to a variety of fish, including brook trout, smallmouth bass, and Atlantic salmon. The village is home to a salmon hatchery and a research center working on reintroducing Atlantic salmon into the fishery.

The byway leaves Route 1 again and heads down Route 191 toward Machiasport. Seaside ledges and bedrock around the bay, including Clark's Point, Birch Point, and Hog Island, contain pictographs known as the Machias Bay Petroglyphs. The carvings were made by shamans of the Passamaquoddy Tribe and some of them are more than three thousand years old.

The road continues and soon Holmes Bay comes into view. At low tide this bay has extensive mudflats that can extend out more than a mile. They are some of the most productive clam flats in Maine and a favorite of local commercial fishermen. From Holmes Bay the road follows a southeasterly path away from the coast and through relatively dense woods until it reaches the ocean again at Little Machias Bay. This detour away from the coast is probably due to the Navy's VLF transmitter located on the Thornton Point Peninsula. The transmitter is one of the largest of its type in the world and allows one-way encrypted communication with American submarines. Needless to say, the area is strictly off-limits.

Left: The town of Jonesport seen from the Beals Island Bridge.

Leaving Little Machias Bay the road brings you to Cutler, one of the prettiest fishing villages along the coast and a great place to take pictures. Neat wooden piers extend into a harbor dotted with lobsterboats. At high tide the water presses up against the road, which runs close to the shore and is bordered on the opposite side by quaint homes and a few commercial buildings — postcard Maine.

From Cutler, the byway continues on to Lubec, one of the most remote sections of the byway, and because much of the land is conservation land, it feels almost deserted. This area is home to the Cutler Coast Preserve, a scenic nature area consisting of a number of natural features, including peat bogs, forest, rocky cliffs, and rocky beaches. There are approximately ten miles of hiking trails and some rustic camping facilities. The trailhead is located off Route 191 and accessed from the parking lot at the Inland Trail trailhead; look for the sign.

Beyond the preserve lands is Lubec, the easternmost town in the United States. It is a quiet town and sits across the Quoddy Narrows from Campobello Island.

Once a mecca for sardines, McCurdy's Smokehouse is the last of seventeen canneries that once operated here. McCurdy's is on the National Register of Historic Places and attempts are under way to preserve it for future generations. Water Street — the main business district — is quiet, with numerous vacant buildings, but there are signs of life at the pier. Fishing still goes on and the boats in the harbor signify the resilience of Maine's coastal communities. One promising venture that might boost the Lubec community is aquaculture. Ocean farming is beginning to take hold in this community and brings hope for the future with it.

Leaving Lubec, Route 189 circles Cobscook Bay on its

Left: Historic buildings in Lubec. Right: Low tide along the Bold Coast Byway.

way to Eastport. This part of the byway is mostly forested, with occasional businesses carved out next to the road. An inlet intersects the roadway every once in a while, reminding you the ocean is not far away. Part of the landscape here is the Moosehorn National Wildlife Refuge. Moosehorn is the easternmost refuge in the Atlantic flyway, so it is an important stop for migrating birds. Consisting of more than 7,200 acres, the primary purpose of the refuge is to protect wildlife, though there are fifty miles of trails for walking, biking, and skiing.

Continuing on you'll pass through the villages of Dennysville, Pembroke, and Pleasant Point. The Passamaquoddy Indian reservation, Sipayik, is located in Pleasant Point and the byway passes directly through the reservation.

Beyond Pleasant Point, the road crosses a narrow strip of land between Cobscook and Passamaquoddy bays, offering grand vistas on both sides of the road. Often at low tide Cobscook Bay completely empties out and becomes one giant mud flat. The Bold Coast Byway ends at Eastport. It has a thriving historic downtown with restaurants, coffee shops, and antiques stores. There is an active waterfront, with a busy state pier and international ferry service. A twelve-foot statue of a bearded fisherman holding a fish stands in the center of town. Often photographed, it was donated to the town in 2001 and is fast becoming an iconic symbol. The statue was a prop for a locally shot Hollywood TV show in which contestants competed for a $100,000 prize. The winner was a New York City firefighter who died on 9/11, one week after the final episode aired.

What better way to end the byway experience than by sitting in one of the local coffee shops, staring out the window as the boats move in and out of the harbor. The only drawback you may find with this wonderful place is that with its proximity to Canada, your electronic devices might switch to Atlantic Standard Time.

Left: Fisherman's Pier in Eastport.
Below: Iconic Eastport fisherman statue.

13

Fish River

13

FISH RIVER

REGION: Aroostook
LENGTH: 37 miles
ROADS: Route 11

TRAVEL TIME: 1 hour
START: Portage
END: Fort Kent

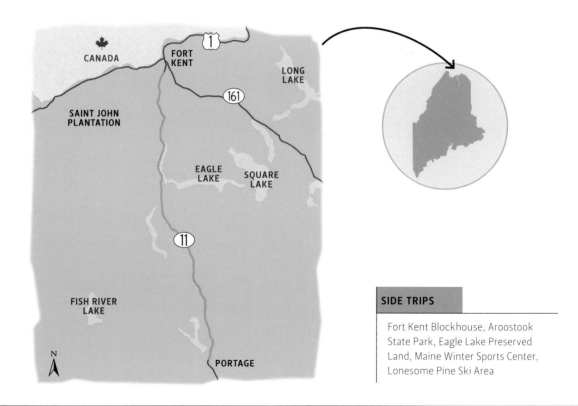

SIDE TRIPS

Fort Kent Blockhouse, Aroostook State Park, Eagle Lake Preserved Land, Maine Winter Sports Center, Lonesome Pine Ski Area

Previous page: Rest area overlooking Long Lake. Right: Portage Lake at sunset. Below: Angling on the aptly named Fish River.

THE 37-MILE-LONG FISH RIVER SCENIC BYWAY (ROUTE 11) IS located in Northern Aroostook County and passes through some of the most rural landscape in America. The byway, buttressed to the west by Maine's North Woods, follows the Fish River and its chain of lakes as it flows north to the town of Fort Kent, eventually emptying into the Saint John River.

The route's southern gateway is the town of Portage, located on the southern tip of Portage Lake. This is an outdoorsman's town for individuals looking for backwoods adventure in its most basic form. With access to one of Maine's deep, cold-water lakes within walking distance of the town center, the location is ideal for fishermen, and trophy fish have been caught in this sportsman's paradise. The eight major lakes that make up the Fish River chain of lakes are Portage, Saint Froid, and Fish River lakes to the west, and Eagle, Square, Cross, Mudd, and Long lakes to the east. The predominant species is wild brook trout and they can be found in almost all of the seven thousand miles of brooks and rivers that make up this system. The gravel bottomed, free flowing rivers, and deep, cold lakes are the perfect hab-

itat for cold-water species of fish. Brook trout, salmon, lake trout, and a variety of lesser known species, most notably cusk, can also be caught in the larger lakes. Cusk is the only freshwater member of the cod family and is usually caught during the ice-fishing season when water temperatures have dropped significantly and the fish are most active. The largest recorded cusk caught in Maine weighed eighteen pounds eight ounces and was caught at Eagle Lake.

In addition to the fishing, the surrounding woods offer great opportunities to hunt deer, bear, and moose. Registered Maine Guides are available in the area, including Hunter's Point Guide Service in Portage and lodging is available year-round at Dean's Motor Lodge or Portage Lakeside Cabins. Other activities include canoeing or kayaking and hiking in summer, and snowshoeing, cross-country skiing, and snowmobiling in winter. The town also has a public beach and nine-hole golf course. The beach offers a quiet place to have a picnic, swim, or just relax; and it's also a great location to watch the sunset. The nine-hole Portage Hills Country Club is a challenging course with hilly terrain and pitched greens, just what you'd expect for a Maine golf course.

From Portage, your trip takes you north to Winterville Plantation, a small farming village with access to Saint Froid Lake. Turning left off onto Quimby Road will bring you to the 2,400-acre lake.

From the village of Winterville the byway continues north through rolling hills, open fields, and woods until it arrives at the town of Eagle Lake, a mecca for outdoor

Left: Beach plums at Square Lake. Right: Farm along the byway. Below: Sunset at Portage Lake.

enthusiasts and the gateway to the 23,000-acre Eagle Lake Preserved Land. The preserve is open to the public for a variety of outdoor activities, including wilderness camping, cross-country skiing, snowshoeing, and hiking. The town is also the site of the Eagle Lake-100 Sled Dog Race. Held each January, the race attracts mushers from all across the United States and Canada. The Eagle Lake area is also a favorite among ATV riders in summer and snowmobilers in winter.

Snowmobiling is very popular in Maine and in no place more than along the Fish River Scenic Byway. The Interconnected Trail System (ITS) maintained by the Maine Snowmobilers Association and the Maine Bureau of Parks and Lands maintains more than 14,000 miles of groomed and marked trails that are open to the public. The Eagle Lake Winter Riders snowmobile club is responsible for grooming seventy-five miles of the ITS from Eagle Lake to Portage Lake. North of Eagle Lake the Fort Kent SnoRiders maintain trails in the Fort Kent area, sections of Saint John Plantation, and northern portions of the ITS along the Fish River Byway.

Potato barn
on Route 11.

An activity gaining popularity in this part of northern Maine is geocaching. Containers of "treasure" are hidden in different places throughout the state and adventure-minded people try to find their locations using clues, secret messages, and GPS devices. There are more than five thousand geocaches in Maine and many of them are hidden in the Eagle Lake and Fort Kent areas, waiting to be discovered by modern-day recreational treasure hunters. Geocaching is a family-oriented activity that can be enjoyed by all ages, adding a new dimension to traditional outdoor activities.

As it continues, Route 11 hugs the shore of Eagle Lake, revealing the beauty of northern Aroostook County. Its panoramic views of distant farmland and rolling acres of potato and wheat fields interspersed with forested hills blend into the contour of the countryside, creating a natural patchwork of beauty. The old farms and potato storage barns that dot the landscape bear witness to the productivity of the fertile soil of the Saint John River Valley.

Once past the town of Eagle Lake, you'll enter Wallagrass and the area known as Soldier Pond. This small town was once home to one of three blockhouses built to defend the border during the Aroostook War. Today it's home to acres of potato fields.

Beyond this section of Route 11 is the town of Fort Kent, the northern terminus of the Fish River Byway. Fort Kent was originally an outpost and the blockhouse built in 1839 — and still standing — was constructed to guard the disputed border between the United States and Canada. The dispute was eventually resolved diplomatically and the "bloodless" conflict known as the Aroostook War faded into history.

Before Europeans arrived, the valley was occupied by tribes of the Algonquin Nation. Later, with the migration of Acadians and French Canadians, the region became a center of Acadian influence. Displaced from Canada, the Acadian people brought their farming heritage to the valley, where it's still in evidence today. Acadian pride is still very strong

in this region and is celebrated through festivals and other social events. The Ploye Festival is one of the largest — a Ploye is a type of buckwheat pancake that was a staple of the region. The Ploye Festival is usually held in conjunction with the annual Fort Kent International Muskie Derby. The Saint John River has hundreds of miles of prime muskie habitat and anglers from around the world come to compete in this event.

Fort Kent is also the home of the 10th Mountain Ski Club and the Maine Winter Sports Center. The club and sports center host local, national, and international events, including a world-class international biathlon competition as part of the world cup circuit. The center has more than twenty miles of groomed trails and is located next door to the Lonesome Pine Ski Area, the northernmost alpine ski area in Maine. This family-oriented ski slope located close to the center of town offers day and night skiing and snowboarding.

The biggest event of the winter season, though, is the Can-Am Crown International Sled Dog Races. Races of 30, 60, and 250 miles start from the town center and run through the North Woods headed for the eventual finish line at the Lonesome Pine Ski Area. Over the course of five days in early March, snow is actually shoveled off the roadsides and onto the streets and the sound of barking dogs and excited mushers can be heard over the voices of thousands of fans. The Can-Am is possibly the most demanding sled dog race in the eastern United States and is a qualifying race for the famed 1,100-mile Iditarod race that takes place in Alaska.

Left: Leaving the starting gate at the Can-Am International sled dog races. Right: Fort Kent Blockhouse.

Saint John Valley

14

SAINT JOHN VALLEY

REGION: Aroostook
LENGTH: 91 miles
ROADS: Routes 161 & 1

TRAVEL TIME: 3 hours
START: Dickey
END: Hamlin

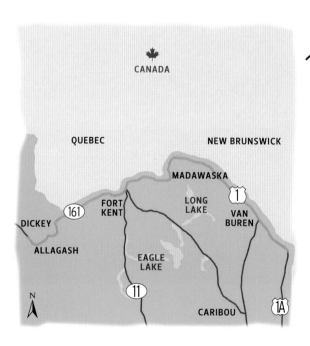

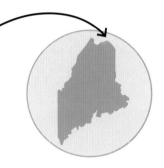

SIDE TRIPS

Fort Kent Block House, Allagash Historical Society Museum, Acadian Village- Living History Site, Allagash Wilderness Waterway, Pelletier- Marquis House, St. Francis Historical Society, Lakeview Restaurant

Previous page: Valley farm and wheat fields. Right: Historic David Daigle potato-seed farm, Fort Kent.

THE 91-MILE-LONG SAINT JOHN VALLEY CULTURAL BYWAY BEGINS at the western end of Route 161 in the small community of Dickey. The area was settled by Scotch-Irish, who immigrated to the valley and found the land to the east already occupied. The land surrounding Dickey is hilly, less fertile, and densely forested, making it more a lumbering community than an agricultural one. Logging began in the early 1800s and supported the shipbuilding and lumber industries, and although less active today, timber is still harvested and contributes significantly to the regional economy.

Starting at the Allagash Historical Society building next to the Dickey Bridge, Route 161 travels east along Maine's northern crown. The town of Allagash is the northern terminus of the Allagash Wilderness Waterway, a 92-mile-long protected waterway that runs through the heart of the north Maine woods. The town itself is built around two thumbs of land and a few small islands created by the convergence of the Saint John River and Allagash Stream. In springtime, when snow is melting and ice-out is imminent, this small hamlet in the woods can be a very interesting place. Allagash is home to a number of guide

services that run canoe trips through the waterway and for many, these trips are viewed as a rite-of-passage.

Driving east from Allagash, you pass through rolling countryside, at times closely following the river. This section of the byway is active forestland criss-crossed by private roads — logging trucks rumbling down the roadway are frequent sites. If you're looking for a spot to stop along this section there's a small park with picnic tables at the town line, it has a pretty stand of birch trees and offers a quiet break to stretch your legs.

At the town of Saint Francis the Saint John River becomes the border between the United States and Canada and the hilly terrain begins to give way to bottomland able to support farming.

Along this section, Christmas-tree farms are interspersed with fields of wheat and potatoes. The randomness of the forested hills playing against the orderly rows of farm trees and alternating fields of potatoes and wheat shows the hands of man and Mother Nature at their best. It also marks the beginning of Acadian and French Canadian influence in the valley.

During the displacement of the Acadian people from their lands by the British, known as Le Grand Dérangement, many Acadians fled to the safety of Quebec. When hostilities ended, families, including some who had married French Canadians, moved into the Saint John Valley. Acadians and French Canadians, connected by a common language and a common religion, began farming the land. The French Catholic heritage is a very strong connecting force and the two most important groups are family and parish. Moving down through the area you can't help noticing the numerous large ornate Catholic churches, usu-

Alternating potato and wheat fields.

ally located at the center of a hub of activity. As a symbol of religious devotion, families have placed crosses in their yard, some of which are quite large. Don't be surprised to turn a corner and find a 35-foot metal cross right in the middle of someone's front yard.

In the early 1900s the Bangor and Aroostook Railroad extended its line to northern Maine and one of its branches ended in the town of Saint Francis. Goods flowed north and potatoes, wheat, and timber moved south, but service ended in the 1990s when transportation by road became more economical. Today the giant turntable built to turn the locomotives around, along with other exhibits, is on display at the Saint Francis Historical Society.

From Saint Francis the byway moves on to Fort Kent, the central hub on this route and unique from a cultural perspective. Walking around, you're as likely to hear French spoken as English. Nearly 50 percent of the inhabitants speak French as their daily language and many people who live in and around the Fort Kent area view the Saint John River less as an international border than as a simple division between two parts of one community.

Passing through town you'll see a granite monument

next to the international bridge to Canada. You should get out of your car here and have someone take your picture — this site marks the starting point of historic Route 1. Stretching 2,450 miles from Maine to Florida, this historic road is the oldest highway on the East Coast and travels through some of the most diverse places in America. If you ever get a chance to explore the Florida Keys Scenic Byway you can get out of your car again and have someone take your picture at that end of Route 1.

Following Route 1 east on Main Street, the byway passes the Fort Kent Blockhouse, built in 1839 to defend the area from attacks from Canada. The soldiers stationed here were also tasked with recovering "stolen" logs as they floated down the Saint John. Continuing over the bridge that spans the Fish River, and past two large Catholic churches, you arrive at Market Street and the old Fort Kent train station. Built in 1902 as part of the Fish River Railroad, the station is now home to the Fort Kent Historical Society. The building is filled with displays and artifacts from when the station was a bustling place as potatoes, timber, and other goods were moved to markets in the south. Behind the station you'll notice very large storage buildings. They once stored barrels of potatoes, but today are mostly vacant, monuments to an earlier time.

Left: Saint Louis Catholic Church in Fort Kent.
Right: Frenchville Historical Society.
Below: Potato grading barrells, Saint John Plantation.

Back on Route 1 you're headed for the town of Frenchville. The byway follows the train tracks and the tracks follow the river, bringing you to a part of the valley taken over by farmland. Family farms were the norm in the nineteenth and early twentieth century, and although many farms have been consolidated, quite a few family farms still exist. In the fall it's not uncommon to see farm stands on the side of the road, many using the honor system, selling sacks of potatoes. Green Mountain and Katahdin are the two most recognizable varieties historically grown in Maine and both contributed significantly to the region's economy. In Frenchville the byway momentarily leaves Route 1 and travels five miles down Route 162 to the town of Saint Agatha. The town is located on Long Lake. Not to be confused with the other twenty-plus Long Lakes in Maine, this is the large deepwater lake of the Fish River Chain of Lakes. Saint Agatha is home to a large number of farms growing wheat and potatoes, and as you drive along you begin to realize some fields run up the hill from the lake in long rectangular strips, broken here and there by rows of trees and woods as windbreaks, while others weave themselves around the land's oddly shaped contours. Even the island in the middle of the lake has a potato farm.

At the head of the lake the road bends to the right and just after the cemetery you'll see the historic Pelletier-Marquis House. Built around 1874, and located today on the Saint Agatha Historical Society property, the house is a classic example of a working-class home of that time period. The historical society also has its museum here and displays many of the tools and equipment that were part of daily life in the eighteenth and early nineteenth centuries. This leg of the byway ends just down the road at the public boat launch. If you brought your canoe or boat to see the byway from a different perspective, then this is the place to go, for everyone else you might want to stop at the Lakeview Restaurant. It has great food, a great view, and parking for cars, snowmobiles, and skis.

Right: Old tractor on a Saint Agatha potato farm. Below: Wheat field in Saint Agatha.

Back in Frenchville, the byway continues along Route 1 to Madawaska, a paper mill town and site of the first landing of Acadians on the southern shore of the Saint John River. A large marble cross commemorates the spot. Agriculture is important and Madawaska has its share of farmland, but the mill built this town. Very large and built at a bend in the river, the Twin Rivers Paper Mill employs a large number of workers from both sides of the border. The town is also home to the Acadian Festival, a weeklong event held each June that celebrates the history of these proud people.

After Madawaska, you pass through the communities of Grand Isle and Lille on the way to Van Buren and the end of the byway, but before the trip is finished there are two places of interest worth visiting. In Grand Isle, Le Musée Centre culturel du Mont-Carmel, a museum and cultural center, is located in the former Our Lady of Mount Carmel Church; and in Van Buren, the Acadian Village, a collection of period buildings, some of which are on the National Register, depict life in northern Maine in the eighteenth and nineteenth centuries. Both museums offer a view into the rich heritage of the people who made northern Maine what it is today.

Acknowledgments

THIS PROJECT WAS COMPLETED OVER A TWO-YEAR PERIOD AND I HAVE MANY PEOPLE to thank. The biggest thank you goes to my family, especially the young ones, Caitie, Jared, Milo, Misty, and Riley, who allowed me to go a-wandering during times when I probably should have been doing something else. I also want to thank Fred Michaud, byway coordinator at the Maine Department of Transportation, for his guidance and fact checking, and my editor, Mike Steere, a patient and supportive person who often uttered the words "Dan, where's the text?" I also want to thank Jennifer Anderson at DOWN EAST for designing the book, and director of photography Dawna Hilton, I know her fingerprints are somewhere on this project. A big thanks to Jim Strang of Katahdin Air; Keith Deschambeault of Rangeley Lake Seaplane Base; and Steve Collins and the boys at Acadia Air Tours, they all helped with the bigger picture; Captain Jamie Robertson of Robertson Sea Tours allowed us to see the byways from a totally different perspective; Eric Sterling, his wife, Mildred, and daughter, Avis, of West Branch Pond Camps, Whit and Maureen Carter of Lakewood Camps, and the great folks at Maynards-in-Maine, they all treated us to unique experiences in the Maine woods; Paul Snell for allowing me the use of his condo as a jumping-off spot on the western mountain trips; Mary Green for making my writing look a lot better than it really is; David and Vickie Lloyd of the Seawall Motel in Southwest Harbor for their hospitality and unlimited supply of early morning coffee; and Jonathan Hines, a 2012 thru-hiker of the Appalachian Trail, who helped us find the missing link. Finally, special thanks first to my friend and excitement junkie, Joe Vizard, who went with me on numerous trips and always kept me laughing; and my teammate and friend, fellow photographer and evolving curmudgeon, Eric Alexander, who went the extra mile, and his wife and kids who understood it worked out to be about an extra ten thousand miles on a road trip through what I consider paradise.